CW00531116

The Tal_ _ger

JUDE

HAPPY ANNIVERSARY

LOVE

DEL

X X X X X X X X X X X

X & 1 FOR LUCK!

Argh
The Tale of a Tiger

M E Buckingham

ASHFORD
Southampton

First published in 1935 by Country Life Limited

This edition published in 1989 by Ashford, 1 Church Road,
Shedfield, Hampshire SO3 2HW

British Library Cataloguing in Publication Data

Buckingham, M. E.
  Argh: the tale of a tiger – (Ashford animal classics)
  Rn: Agnes Mary Easton
  I. Title
  823'.912[F]

  ISBN 1–85253–159–2

Printed by Hartnolls Limited, Bodmin, Cornwall, England

Contents

	Foreword	vi
	Prologue	1
I.	Argh's Mother	9
II.	Back to the Jungle	17
III.	Freedom	25
IV.	Flight	33
V.	Sanctuary	40
VI.	The God of the Jungle	45
VII.	A Pilgrimage	50
VIII.	The Coming of Argh	55
IX.	First Lessons	60
X.	Man	67
XI.	Fire!	72
XII.	Rescue	79
XIII.	The Taming of Argh	83
XIV.	Fame	92
XV.	Fever	97
XVI.	Stolen—But Paid For	102
XVII.	Through the Jungle	107
XVIII.	Captivity	113
XIX.	Trial by Jury	120
XX.	Valhalla	127
	Epilogue	135

Foreword

Beneath the surface of this poignant if rather sentimental tale, set in the twilight years of the British Raj in India, there is another deeper and more tragic story, unstated but nevertheless there to be read. A story not simply of the trust between a man and an abandoned tiger cub and its eventual betrayal, but of the age-old conflict between civilisation and the wild, between the animal kingdom, dumb in its endurance, and man—man the hunter-turned-farmer, man the restless, 'hairless monkey', man the author of change. It is a perennial conflict which continues even to this day, indeed has multiplied and intensified even to the point where the complete extinction of many previously numerous species is imminent. Throughout the world man is pushing back the frontiers of his own kingdom, endangering the existence of many familiar if now statistically insignificant species, not least the African elephant, the rhinocerous, the mountain gorilla and the blue whale.

Although the underlying cause of this depletion is the

same now as it was in the inter-war years when M E
Buckingham was writing—that is, the destruction or pol-
lution of these animals' natural habitats—there is nevertheless a less forgivable though perhaps no less inexorable
cause—man's entrapment and abuse of animals for his own
vanity or pleasure: rhinocerous horns for 'aphrodisiac'
potions, elephant tusks for ivory, whale fat for cosmetics,
tiger skins for floor coverings, circuses and zoos for idle
curiosity and entertainment (with the honourable exception
of some zoos which now take the lead in the conservation of
species).

In this remarkable story it is possible, with a short stretch
of the imagination, to empathize with the excitement and
pleasure the Maharajah's subjects take in the capture and
torture of Argh's mother, since it is their lot to compete with
the tigers over the soil reclaimed from the jungle. Yet it is
with the caged 'beast' that our real sympathies lie. Her
native captors torment her cruelly, but this is no racially-
motivated slur on the part of the author, this is simply nature
'red in tooth and claw'. The tigress, having escaped from
her cage, feels no compunction in swatting like flies all those
in her path, and the cruelty she has suffered is passed on to
the next generation when her own cub, Argh, is first rescued
from fire by the colonial forest officer and then tricked into
captivity by a circus owner from England. Will the cycle of
mistrust and betrayal ever be broken?

The answer of course lies with man, capable of great kindness, great cruelty and, at best, great understanding. Perhaps Mrs Buckingham divines some hope in the maturing of the Empire and its officers, in the man who is civilized yet wise in the ways of the jungle, the poacher turned gamekeeper. It is not for her to know that the Empire and its ethos were even then in terminal decline, but her upright and kindly hero is indeed prophetic in his adoption of the camera rather than the gun.

Argh's tale is not a happy one, nor is it free from contemporary opinion in its incidental exposition of the natural history of the Bengal tiger, but it is an imaginative and thought-provoking slice of colonial culture, a worthy jewel in the already glittering crown of animal tales from the old Raj.

Prologue

IT WAS five o'clock on a September morning. The sun had not yet risen and the world was still quite dark. All the flowers in the Manor House garden were sound asleep ; the dogs were lying in their baskets in the wash-house like stuffed images, their toes tucked round to keep off the draughts of the early morning, dreaming of the rabbits that they had chased the day before. They could not hear the snores of Mrs Power the cook, who slept in the room above with their implacable enemy, Thomas the cat, curled up at her feet. In fact the Manor House and all its inmates were asleep, at least all except one.

Jeremy Mainwaring, or Jeremy the Second, was wide awake. He was perfectly sure that it was time to get up, and the fact that it was dark filled his soul with despair. It must be that it was raining cats and dogs for it to be so dark, and it was long before he dared slip out of his bed and look out of the window. At last he summoned up his courage, tiptoed across the floor, and pressed his nose against the glass. No, it was certainly not rain that made everything so dark !

He drew back from the window and listened intently. There was a complete absence of ' getting up ' noises : the only sounds that he could hear were the deep puffings of Nanny, who slept in the next room with his young brother. It dawned on Jeremy that after all he must have got up in the middle of the night. He went to the mantelpiece and tried to see the clock, but it was too dark to make out the face. He dared not turn on the light, as Nanny always woke up if the light shone through the chink of the door that was left open between Jeremy's room and the nursery ; so he tiptoed back to his bed.

' It must be the middle of the night,' he thought. ' And the middle of the night is twelve o'clock, and if Uncle Jeremy doesn't get up till eight—then he has to dress—what a long time grown-ups take to dress—and breakfast—another long, long time, say two hours at least—and reading his letters *and* the paper—'

Jeremy sighed despairingly and tried to work out the number of hours that he would have to wait before his uncle would be ready to face the world, but the effort was too much for him and he dropped off to sleep.

Uncle Jeremy, or Jeremy the First, had arrived the day before. He was a terribly exciting person, for he had just come all the way from India and was full of the most exciting adventures. Jeremy and James had listened with their eyes popping out of their heads when their uncle told them story after story of all the wild beasts he had met, all personal friends of his.

They swelled with pride as they listened. Many of their friends had uncles who shot big game—skins of tigers and bears would cover their floors, while great horned beasts frowned from the walls as proof of such prowess with the gun—but Uncle Jeremy was different. He had sheaf after sheaf of photographs instead of skins, photographs which he had taken of the animals in their own jungles, as well as those which roamed about his bungalow, free to go and come as they pleased, but preferred to stay with him because they liked him so much.

Just as Nanny had come in to whisk them off to bed Uncle Jeremy had promised that, if the next day were fine, he would take them both to Whipsnade to see the tiger that had lived with him for nearly three years. It was of this tiger that Jeremy the Second dreamed when, worn out by his efforts at mental arithmetic, he had fallen asleep once more.

In his dream he was walking down the village street when suddenly there was a cry of ' Fire ! Fire ! ' Jeremy quickened his step, and on turning the corner by the sweetshop was a little surprised to find that the rest of the village had disappeared.

Instead of cottages with neat little gardens laid out in front of them there was a thick forest of trees with huge flames licking up their trunks. The village fire-engine was there, but curiously enough it had dwindled to the size of a doll's perambulator, and on looking at it closely he saw that instead of a fire-engine it was the circular spray which watered the Manor House tennis-lawn during dry weather.

Jeremy joined the group of village people who were watching the fire, and even while he was watching, the fire turned into a Catherine Wheel. Round and round spun the Catherine Wheel, shooting its sparks in all directions. Slowly the wheel died down until there were only two sparks left. Then, instead of going out, they grew larger, until Jeremy saw that they were the eyes of a huge tiger.

'It's Argh! It's Argh!' he cried excitedly. 'Argh! Argh! Argh!'

'And what might "Argh" be, Master Jeremy?'

Jeremy sat up and rubbed his eyes. 'Oh Nanny, I thought you were a tiger!' he explained sleepily.

'But that's no reason why you should call me "Argh" or some such outlandish name!' answered Nanny, as she pulled back the bedclothes. 'Wherever did you get it from?'

'It's Uncle Jeremy's tiger that was called "Argh", Nanny. It means "fire", you know. He was called that because he came out of a fire. Now do you understand?'

But Nanny didn't understand at all. 'Tut, tut!' she muttered to herself. 'Tigers coming out of a fire indeed. Whoever heard of a tiger coming out of a fire? Filling the children's minds with such nonsense!'

But Jeremy and James didn't think that it was nonsense as they spun along the road, squeezed into the front of Uncle Jeremy's two-seater. His behaviour had been splendid that morning, for he had refused a second helping of mushrooms and bacon rather than keep his nephews waiting a second longer than was necessary, and they had actually started before there was time to ask 'Oughtn't we to go now?' more than twice—a truly marvellous uncle.

For miles the car threaded its way through leafy lanes.

Uncle Jeremy was fond of what he termed 'short cuts ' and, after a long study of the map on the night before, he had planned their route most carefully so as to avoid the main roads. The sun, which early that morning had seemed blotted out for ever, now shone brilliantly and James, who was much moved by the prospect of the joys awaiting them, burst into song.

For once Jeremy allowed him to sing without interruption. As a rule James's songs moved him to fury, but he remembered that singing is hungry work and that, if allowed to continue, James would produce some kind of provender. Jeremy had not been mistaken. Before he had got to the end of the fourth verse of ' the raging seas do roar ' James extracted a paper parcel from his overcoat pocket and opened it carefully.

' Will you have a sandwich or a piece of chocolate, Uncle Jeremy ? ' he asked politely.

Uncle Jeremy regarded, out of the corner of his eye, the emergency ration which Nanny had provided in case an uncle so lately come from the wilds of India might forget to give his nephews enough food. The sandwiches, which doubtless had been excellent when they had left Nanny's hands, had not been improved by their sojourn in James's pocket, and Uncle Jeremy declined the offer with thanks.

Jeremy heaved a sigh of relief, for he was not sure how much an Uncle of that sort might not want to eat. In a moment the two boys were munching solidly, while the car purred steadily along the Hertfordshire lanes. As James crammed the last bite

into his mouth they reached a main road with a huge yellow
sign-post standing up in a green triangle of long grass.

' Zoo ! ' shrieked Jeremy, who was eight and a half, and
could read.

' Zoo ! ' yelled James, who was six and couldn't—at any
rate not at that pace—though he liked to pretend that he could.

They both expected that Whipsnade would appear round
the next bend in the road, and it was a blow to find that this
was far from being the case, but they drew comfort from the
fact that so magnificent a yellow sign-post could not possibly be
entirely wrong. When at last a humbler white sign-post confirmed what the yellow one had said they knew that their destination must be close at hand. The car turned and twisted down
a road so narrow that two cars could hardly pass each other, and
at last they found themselves in front of the huge car park outside
Whipsnade.

' Whipsnade ! Whipsnade ! ' shouted the two boys.

' Here, here, you two ! ' remonstrated Uncle Jeremy. ' If
you make such a din they will put you into the monkey-house by
mistake, and I shall have to visit Argh by myself ! '

A horrified silence greeted this remark until Jeremy, glancing
at his uncle's face, decided that he had meant to be funny.
' It's all right, James ; Uncle Jeremy is being funny,' he re-
assured his brother.

' It's a temptation that few uncles can resist,' Uncle Jeremy
murmured apologetically.

' That's quite all right,' remarked Jeremy, ' we're used to it. There ! There's the gate ! Hurry, Uncle Jeremy.'

At last the car was parked, the turnstile had clicked behind them, and they stood in the wide sweep inside the gates. James gazed at the long wide road in despair. The thought of spending precious hours tramping along it appalled him, though nothing short of torture would have made him admit the fact. To his relief Uncle Jeremy strode straight over to the bus which stood waiting at the side of the road, and they clambered in hastily. It dawdled along the road in the most delightful way, so that its passengers could have a good look at the animals as they went past.

' Are they really *wild*, and *fierce* ? ' enquired James a little doubtfully. ' They look so gentle.'

' They wouldn't be gentle if you were to get over the railings and chase them ! ' answered Uncle Jeremy.

' Those—those beasts over there look rather like cows with fur coats on,' said James.

' Cows indeed ! Those are bison, you ass ! ' scoffed his elder brother. ' Don't you remember the story I read to you— oh ! oh ! Zebra ! ' He broke off with excitement. ' There ! In front of us, *and* they are talking to people through the railings ! '

But James had turned his attention to the other window. ' Polar bears ! *Real* polar bears ! ' he said, in a hushed awe-struck voice. ' And I *think* I saw a little bit of tiger but I'm not sure.'

' Well, this is where we get out,' said Uncle Jeremy as the bus stopped. ' We will walk along and see the zebras later. There are your polar bears, James, and below, in that hollow, unless I am very much mistaken, we shall find my old friend Argh ! '

The two boys pulled their uncle down the little hill that led to the front of the tiger-pit at a pace ill-suited to his years, and all three were slightly puffed when they reached the railings. For a moment both boys stood quite still and gazed in awe at the huge tiger that lay on the rock below them.

' Is it—is he Argh ? ' asked Jeremy anxiously.

' You will soon see,' answered his uncle.

He placed his arms on the railings, and, leaning slightly forward, spoke to the somnolent tiger quite softly. ' Argh ! Argh, old man ! Don't you know me ? '

The tiger lifted his head sharply and pricked his ears. Slowly he rose to his feet, dropped down the terraced rock, and stood below the railings. For a moment he stood stock-still, only his head moving ever so slowly from left to right. Then, placing his great paws against the wall of the pit he stood up, eight gleaming yellow-and-black feet of him, and roared a welcome to his old master and friend.

CHAPTER I

Argh's Mother

THE MARRIAGE festivities of the Maharajah of Chandupur had been in full swing for two nights and days. The greater part of the population were by now gloriously drunk, but they carried on with their singing and dancing until they toppled over asleep. Nobody took any notice of the sleepers : the rest of the merry-makers, squatting in circles on the ground, clapping their hands in time to the beat of the tom-toms, or staggering into the middle of the ring, flung themselves about in a frenzy of excitement until they collapsed from sheer exhaustion.

On this, the third and last evening of the ceremonies, even the strength of the most enthusiastic was beginning to flag, and it was clear that some special attraction must be staged at once, or else the marriage feast would fizzle out like a dying firework. The Maharajah too, glittering in his wedding finery, had for some time been showing signs of boredom, and although he no longer held the power of life and death over his subjects, owing to the squeamishness of the British Raj, he could still make life unendurable for them—and thoroughly enjoyed doing so on the least provocation.

The entire staff was trembling in anxiety and bewilderment —it was no joke to be a servant of the Maharajah of Chandupur when things were not going well. Some scratched their heads, others wrung their hands, while the Dewan himself wept out loud, for as Prime Minister he had more to fear than any of the others. What *could* he do to amuse his master? Pariah dogs had fought and died for the Maharajah's entertainment: hooded cobras, summoned from their baskets, had swayed to the weird drone of the pipes: young pythons had coiled themselves around their owners' necks and bodies: nautch girls had danced before him until they fell senseless at his feet, only to be dragged away and replaced by a fresh supply.

' There is yet one box of English fireworks, not yet exploded. Shall we light those?' suggested one man, not very hopefully. All the best fireworks had been used long before, and the box he had discovered only contained two dozen giant squibs. Even if they did succeed in pleasing the Maharajah it could not be for more than a few minutes—as the other officials were quick to point out.

Suddenly the Dewan forgot his tears and uttered a yelp of triumph. He had had a brain-wave of such brilliance that it brought a sparkle of excitement to the tiredest in that weary court.

' The tigress!' he shouted. ' The tigress! Cause the cage to be wheeled before the palace and let the fireworks off inside it!'

His idea was greeted with a yell of enthusiasm : all the officials and servants started to gabble at the top of their voices. They seized the fireworks and rushed towards the courtyard behind the palace where the Maharajah kept his menagerie. Other voices took up the shout until the mud and plaster buildings seemed to rock with the noise of it. ' The tigress ! The tigress ! '

The only man who did not hear the uproar was the man in charge of the tigress. He lay wrapped in a drunken sleep that little short of the last trump could break. But nobody paid any attention to him—there were plenty of willing hands to drag the heavy cage to the front of the palace and place it where the Maharajah and his friends could see it from the draped platform where they were sitting.

The little state of Chandupur is mostly jungle. The people in the villages cultivate just enough for their bare needs, waging a never-ending warfare against those beasts in the jungle whose delight it is to devour young crops. Their chief source of income is drawn from the skins they take from the animals that creep down at night to steal their paddy, and from the tigers and leopards who have come after their herds and have been shot by the shikaris. Sometimes they capture a living animal, when the beast is either kept by the Maharajah, sold to a zoo, or killed for its skin.

The tigress now being drawn towards the centre of the marriage feast had been caught six months before in a pit dug for the purpose. She had been a beautiful beast when they had brought her in ; her coat was perfect, her eyes bright, and her temper as savage as anybody could wish. Now after six months of close confinement, cruel treatment and underfeeding, her spirit had been broken ; her lovely amber eyes had lost their lustre and her coat looked like a doormat ravaged with moth.

She had given the Maharajah great pleasure when he had first seen her, for at a shout she would lash herself into a fury, and when iron rods were poked into her through the bars of the cage from all four sides at once she would fight like a demon, and had ripped up the arm and shoulder of one of her tormentors who had come too near the bars.

' A very, very fine beast,' the Maharajah had cried ecstatically. ' Take that fool away and give her another poke ! '

For a week the tigress became the seventh wonder of the universe, being visited two or three times a day by the Maharajah and his friends. Then, like all her predecessors, she ceased to be a novelty and was left neglected in the menagerie. With neglect also her fire died away ; she became listless and lay sulkily in the cage, even when some of the servants prodded her in fun. She would barely raise her head to growl even when the sharp points of the rods found a tender spot in her bony body, and they soon tired of tormenting her. The tigress knew that her chance would come one day, and until that day should come she bided her time and nursed what strength cruelty and starvation had left to her.

In the meantime the Maharajah eagerly awaited the arrival

of the cage. The plan for his amusement had been explained to him and he was delighted that once more he would have a chance of seeing his old favourite rage helplessly in her cage. She had proved such a dull amusement lately that he had almost forgotten her existence.

The courtyard of the palace was crowded with people as the cage was trundled into position, for a rumour of what was to come had quickly spread, and many a sleeper had been awakened and pulled tottering to his feet. A space had been cleared on one side of the square, so that the Maharajah's view should not be blotted out by the seething masses who were wrestling with each other in their anxiety not to miss a moment of so splendid a spectacle.

The first squib proved to be a complete failure and would not light, but the second was shoved into the cage as soon as the match had been applied and would no doubt have done its duty if the tigress had not stretched out a huge paw and extinguished it before it had time to start spluttering.

These two failures worked the crowd's excitement up to fever pitch, and, as the third squib was thrust into the cage just as it began to crackle, a roar of cheering broke from the people, who had now pushed forward in their excitement and were packed round the cage. The Maharajah, his dignity forgotten, stood up and shouted with the rest, dancing up and down on the platform and waving his fat hands in encouragement— whether of the fireworks or the tigress it was not quite clear.

As the squib landed spitting on the floor of the cage the tigress leapt to her feet.

' Put in another ! Put in another ! ' yelled the crowd, but the man with the fireworks was too overcome by the success of the last one to bother to light another.

The tigress seemed to have gone mad with terror and rage. Never had she given so magnificent a show. Round and round the cage she went, striking at the squib, which always darted away just as her paw descended—and then the squib, with a last fizz, went out.

' Another ! Quick ! Hi ! You there ! Light another ! ' yelled the Maharajah, but his voice was drowned by the shouts of the crowd. But the box was not to be found, for it had been kicked further and further back as the crowd surged forward, and was now being used as a footstool by a man who wanted a better view.

By now the light had gone and yells for torches added to the din. At last somebody found the box, pushed the man off his footstool, and brought it triumphantly back to the side of the cage.

' Open the cage and put the whole box in ! ' shouted a voice in the crowd.

' No ! No ! The tigress will get out ! ' screamed several men who were sober enough to realize the danger.

The tumult round the cage increased. Inside, the tigress, lashing her tail with fury, her eyes blazing with hate and her teeth gleaming in the light of the torches, tried to guard all sides of the cage at once.

Then one man, drunk with malwa beer and excitement, raised his hand to the padlock which held the door in place. The tigress leapt at him, her claws tearing the flesh from his knuckles. He dragged his hand back with a scream of pain and as he did so dropped his torch, which he held in the other hand, into the box of fireworks.

A second later squibs were exploding all over the place. Maddened by the pain in his hand the man tried to bolt through the crowd, away from the cage and the exploding fireworks. His panic infected the crowd, who thought that the tigress must have broken loose, and the courtyard was filled with screams of terror. Men trampled on each other in their anxiety to escape from they knew not what : the Maharajah was hustled into the palace, and all doors were shut and bolted.

Once more the tigress, terrified out of her reason, leapt at the door of the cage. In her frenzy her strength was doubled, and three times she hurled her weight against the lock. The chain of the padlock snapped—she was free !

In such a moment of panic her one thought was to get back to the jungle. The crowd were already flying to right and left in their terror, and down the path that was opening up between each half of them the tigress made her way over the bodies of men who had been knocked down in their flight, treading on them, but not attempting to maul them, although her great claws tore gashes in their flesh as she pushed them to one side. Those who could not get out of her way quickly enough were less lucky, for at the least sign of a blockage in her path to freedom the tigress struck out—great blows from the shoulder which crushed the life out of her victim in a second.

At last she fought her way clear of the palace-yard and was racing down the street ! The news of her coming had spread like wildfire : but for a few fugitives hastily concealing themselves inside the huts that lined the narrow road she found herself in sole possession of the town of Chandupur.

On and on she raced until the thick jungle swallowed her up and she could lie hidden once more in its green peacefulness and regain her health and strength.

CHAPTER II

Back to the Jungle

ONCE MORE in the jungle with the soft earth beneath her feet, and a vast forest of trees and thickets in which to hide, the tigress dropped to the ground. Her rage and terror had subsided and she found herself overcome by such a weariness as she had never experienced before her capture.

All that night she lay in a patch of long, lush grass—licking her wounds and sore paws, watching the dancing lights of the town where she had suffered so much disappear one by one, ready to sneak into the heart of the jungle at the first sound or scent of her hated enemy—man. As soon as the grey light of approaching dawn began to streak the sky she rose painfully to her feet. Instinct told her that she could not stay where she was during the long hours of daylight, for she was in no condition to risk being pursued by her enemies, should they happen to find her.

The tigress hadn't the strength for a long journey to a safer neighbourhood, so she made her way back to the ravine in which she had lived before her capture. Every step was painful, for in bursting out of the cage one of the bars had cut her shoulder nearly to the bone—but she was free, and the courage that is in all wild things gave her strength to go on.

She was careful to avoid the path which had led to the pit in which she had been trapped. It was just as well that she did, for the nine-foot hole had been covered once more with its treacherous roof of twigs and leaves, and in her present worn-out condition her wits were not sufficiently alert to prevent her from blundering into it again.

At last she reached the mouth of the ravine and slowly dragged herself through the jungle which clothed its sides, until, just as the sun began to make itself felt through the thick leaves, she reached sanctuary.

The cave was situated at the very head of the ravine, its mouth being screened by heavy trees. A little below it there was a beautiful patch of green turf, and near at hand ran a stream which now flowed peacefully in its bed, although during the rains it became a roaring torrent.

The cave had been uninhabited for many months, ever since the tiger, unaware of the tragedy that had overtaken his mate, had returned to their lair one morning to find his home deserted. His deep roar had echoed and re-echoed up and down the ravine, making the jungle-folk hidden in its sheltering trees tremble for their lives.

' What has become of you ? What do you mean by staying out so late ? ' he seemed to say. ' How dare you treat me, the biggest tiger in the jungle, with so little respect ? Come home at once ! '

Birds had cried shrill warnings to each other ; monkeys, swinging in the tree-tops, had chattered hysterically ; while the crows had swirled above him, informing all and sundry that below them was a tiger—a fact that they were fully aware of already. Again and again his mighty roar had brought terror into the hearts of the weaker jungle-folk, but his only answer had been the echo of his own voice, reverberating through the ravine. At last, having failed to get any answer to his summons, he had entered the cave and slept through the day. He returned twice more to the cave, but game was getting scarcer and scarcer, so he departed for good to find a better hunting-ground further afield.

On that fatal morning the tigress had killed early—a fine fat spotted deer, and after satisfying her own hunger and tending the cubs, she had summoned her lord and master to the feast. Then she sat down and cleaned herself and the cubs.

By the time the tiger had arrived there was not a hair out of place on her gleaming hide ; the black bars shone like ebony on her golden coat. The cubs, now nearly a year old, gambolled round their mother in the moonlight and, but for the gruesome remains of the deer, she might have been a gentle domestic cat playing with her kittens.

As the tiger approached the cubs drew back. They had a wholesome respect for their father, and knew from experience that it was unwise to attempt to play with him before he had eaten his fill. The tigress rose to her feet and the two animals greeted each other by twitching back their lips and uttering a low growl ; then the tigress turned and disappeared into the jungle in search of water, the cubs trotting at her heels, for after a meal a tiger must drink, and drink deeply.

The nearest water was the stream which flowed down the ravine, and when she had finished drinking the tigress turned her head towards home. But she was destined not to see the cave again for many a long day.

There was nothing to show that the path in front of her was in any way different from the path immediately behind. A certain darker patch of colour was obliterated by the moonbeams slanting across it through the trees. One moment the tigress was striding along the path ; the next the ground had opened beneath her hind legs, and she was struggling wildly for a foothold, scarring the mouth of the pit with deep claw marks in her frantic effort to save herself. For a moment it seemed that she might drag herself over the edge to freedom, but her precarious hold suddenly gave way and she fell to the bottom with a crash, carrying in her wake a great pile of loose earth from the lip of the pit.

Again and again she had tried to spring from the bottom, but there was no room for her to take off, and each time she fell back. At last, wearied and bruised, she lay down at the bottom of the hole, awaiting her fate. The noise of her struggles had awakened the forest ; monkeys chattered, parrots squawked ; two young wild boars, having thrust their astonished noses through a thicket to make sure of their enemy's downfall, rushed off to carry the news to the rest of the sounder.

Meanwhile the cubs ran round and round the mouth of the pit, whining encouragement to their mother, longing to run away from this unknown terror, and yet not daring to leave. At last they lay down also, with their heads on their paws, staring fearfully into the depths of this strange cave.

They had not long to wait, for a herdsman had heard the tigress's roars of rage and terror, and had hastened to the village with the news. Being almost certain that success had crowned his efforts the chief hunter—or *shikari*—had set out for his trap with a band of helpers, a bullock-cart, several picks and shovels, and two long, heavy poles.

Every man who owned a gun took it with him, and soon the little army was threading along the narrow track which led to the jungle. The herdsman, as a reward for having brought the news, was allowed to accompany them, and he trudged along the dusty track beside the bullock-cart in which the *shikari* sat alone in his glory.

' It will be a tiger of great size,' he assured the *shikari*, ' for its roar filled the jungle all night. Your great cleverness in trapping such a monster will bring you into great favour with the Maharajah, oh, most mighty hunter ! '

The *shikari* glanced anxiously at the tiny wooden cage towed behind the bullock-cart. ' Will it stand the attack of so mighty a tiger?' he wondered uncomfortably, clutching his gun more tightly.

As the band of men approached the trap the cubs turned to flee, but they were too late. The *shikari's* eldest son, a hot-headed youth and vastly proud of his marksmanship, shot one dead, and his nearest neighbour, not to be outdone, accounted for the other : the cubs, their troubles over now, lay twitching at the mouth of the pit. It was the best thing that could have happened to them, for they were not big enough to hunt for themselves, and had they escaped they would soon have died of hunger, even if they had escaped a more violent death.

But the *shikari* was furious. He had already been reckoning how much they would fetch when captured, and now all he could get for them was the price of their skins, which would be small, for the price is regulated by the size of each pelt. He rated the two men long and noisily, and then, his stock of abuse running out, he stepped cautiously to the lip of the trap and looked down. Immediately the tigress sprang at him and he stepped back hurriedly.

The cubs had already shaken his faith in the herdsman's theory, and one glance confirmed his suspicion, but nevertheless there was a tinge of relief in his disappointment. The animal in the trap might not be a tiger, but it was certainly a very fine tigress. He turned to the men behind him.

' Get to work ! Get to work ! ' he cried, and the men with picks and shovels started to dig a trench leading towards the trap.

By midday only a thin wall of earth separated the trench from the trap. The wheeled cage was shoved down the trench until it touched the side of the hole in which the unhappy tigress lay.

'Lift the door and break down the earth,' ordered the *shikari*, who was fidgeting with excitement now that the tigress was so soon to be his.

Quickly the heavy poles were brought forward and pushed between the bars of the cage. 'Aye, aye, umph,' droned the men in chorus, as they drove the poles against the thin partition of earth that remained.

The tigress leapt to her feet as the wall of earth tumbled down ; ahead of her she saw light—nothing but a few puny men stood between her and freedom ! She leapt out of the trap and into the cage at one bound—but as she landed in the cage the doors dropped and she was confined even more closely than before.

For the next five minutes there was a mad whirl of paws and claws and huge snapping teeth. The cage rocked and groaned : the tigress roared and snarled—nothing, it seemed, could withstand such a terrible onslaught. And yet the stout young tree trunks from which the cage had been made held good—the trap doors, rattled and shaken in the tigress's teeth, did not shift from their slots. If the cage withstood that first terrible onslaught there would be no fear of its collapsing afterwards during the journey to the village.

Racked with thirst, rage and terror the tigress exerted all her strength, but the men who had built the cage had done their work thoroughly—it rocked upon its wheels as though shaken by an earthquake, yet, tear as she might, the heavy log bars stood firm until even her stout heart had at last to acknowledge defeat, and she sank down on the floor of the cage.

All the way to the village she had watched for a chance of escape, and never during all those weary months which were to follow did she relax her vigilance, and at last she had reaped her reward.

CHAPTER III

Freedom

SHE WAS free ! Before seeking the shelter of the cave the tigress drank deeply from the stream ; then she swam about in the water and rolled over and over, partly to ease the stinging pain in her shoulder and paws, and partly to remove the foul smell of captivity that clung to her fur.

When she had rolled over twice, and had known the comfort of the cool water washing her wounds, she dragged herself out of the water and cleaned herself carefully. In spite of the handicap of her wounded shoulder she managed to smarten up her appearance and get rid of the dingy film of dirt that she had collected in her prison. The flies started to torment her, settling on her wound, so once more she rolled, this time in the soft earth at the edge of the stream, until it was covered with a warm mud plaster.

When all her instincts for cleanliness and protection had been satisfied she climbed up to the cave. She approached the entrance warily, but her nose soon told her that her late home had been untenanted ever since her capture. Gratefully she crept into its cool darkness and sniffed carefully into each nook and cranny. When all her suspicions had been allayed she lay down at the very back and slept.

For two days and nights she made no attempt to leave her refuge. Fever, caused by her wound, raced through her blood, and all that she needed was peace and the water that flowed at her door.

By the third night the fever grew less and she became conscious of the fact that she was hungry, but she knew that she was in no condition to pull down game. Even when she sighted four Sambur hinds at the edge of the forest she could not muster up strength enough to stalk them for long hours through the jungle. She was forced to descend to that lowest of tiger's diet—frogs. Although a poor substitute for a meal, they were better than nothing, but at times she longed for the lumps of goat's meat which had been her sole food for the last six months, even though they smelt strongly of their contact with human hands.

At last, on the tenth night after her escape, she decided that the time had come to take up the trail once more. Her shoulder wound had healed, and the pads of her feet were no longer tender, but her muscles were stiff after her captivity and she moved without her accustomed lissomness. Her striped coat hung limply on her gaunt frame and across her shoulder ran an ugly scar, from which the fur had receded, leaving a bare patch.

She left the cave and for a few seconds stood on the grass patch, testing every breeze in the hope of it bringing her news of the whereabouts of a meal. Then she stretched herself and yawned.

All around her the jungle lay wrapt in silent darkness, for the moon had not yet risen. Again she sniffed the air and her eyes glowed like hot coals. Somewhere among those black tree-trunks a dinner lay hidden, and the tigress meant to dine that night if it cost her the last ounce of her strength. She stepped delicately across the grass and disappeared into the forest without a sound.

Deep into the jungle the tigress plunged, invisible, and silent as a shadow, her whole being concentrated on her desire for food ; yet, hungry as she was, she did not neglect any of the precautions learnt from long experience of the jungle and its ways.

There was no haste in her movements. She slipped into the cover of a thick bush and crouched down. There was nothing but the burning fury of her eyes to show that she had any thought in her mind other than the finding of a peaceful retreat.

One hour, then another, passed. Still no luck came to the starving tigress. A sounder of pig passed her hiding-place, but unfortunately they had come up to windward of her, and, scenting her presence, had plunged into the forest before she could get near them, alarming the jungle with the crashing din of their flight, and squealing a warning of the danger that was lurking in that thicket.

Another weary hour passed and the moon rose. The light flickered through the trees, making grotesque shadows which, to the eyes of the starving tigress hidden in the undergrowth, seemed to take the form of fat pigs and juicy herds of deer. But she did not trust to her eyesight, and her nose told her that these shapes that flickered and danced before her were unreal, and she lay still.

At last, down the wind came a smell that made the tip of her tail quiver with excitement : a cheetal hind suddenly appeared in a patch of moonlight. A herd followed, on their way to drink at the stream, and scenting no danger, moved briskly towards the opening of the ravine. Instinct told the tigress that they were making for a gully high up above the cave, where the grass grew soft and green, and where, even in the hot weather, there was a trickle of water in the stream.

She rose quietly from her lair and made a wide detour, always keeping to leeward of the herd. Slowly she followed them, hiding in a thicket whenever they were crossing an open space, and making up on them as soon as they had vanished into the cover of the trees.

Her instinct had not betrayed her ; in a few minutes the herd had reached the ravine and emerged from the shadow of the jungle. Now was her chance : she crept swiftly forward inside the verge of the jungle, pushed on a little beyond them while they stopped to graze, and finally crouched at the side of the track, ready to make her kill when the moment should arise.

She let the main herd pass, her eyes never leaving a fine

buck who hung back until the herd had dropped over the lip of the gully before he also stepped out into the track and started to move slowly in their wake.

The tigress crouched lower and lower, her body flat on the ground and her claws gripping the earth. Every now and again she shifted her position slightly, adjusting her balance for the perfect spring, and as the buck stepped delicately past her she shot through the air with a roar that shattered the silence of the jungle, and landed on top of him, burying her teeth and claws in his back.

As she landed she jerked the buck's head backwards, breaking his neck. The next moment she and her victim were rolling together down the side of the gully.

At the first sound of that charge the herd had scattered. She could hear them crashing in all directions through the forest. The sounds slowly died away and she found herself alone with her kill, the blood of which was already staining the clear water of the stream.

She was bruised and shaken by her fall, and the forest rang with her roars as she pulled the buck out of the water. If her strength had been normal she would have dragged the carcase as far as the cover, but it was all she could do to get it clear of the stream.

While all the jungle folk trembled, she squatted down to her first real meal for over six months. With furious growls she tore the flesh from the haunches of the buck as though she were wrestling with a living enemy. Hair, skin and bones still warm with life, she tore from the quivering carcase in huge mouthfuls, until, her hunger satisfied, she moved down to the stream to quench her thirst. She drank deeply, dyeing the water red with the blood of the buck, and gradually her rage subsided and the fire died out of her eyes.

When she had drunk her fill she raised her head from the stream and sighed deeply, as a child sighs after drinking—as she did so her eyes caught sight of a shape moving in the bushes at the top of the gully.

The tigress looked round her slowly, and then stood rooted to the ground. She knew that shadow in the bushes was another tiger, and to reveal her presence was almost sure to force a fight. Again she moved her head from side to side looking for a way of retreat—she was heavy with the meal that she had just eaten and she had no wish to fight, although it would mean giving up the remains of the buck which lay a few yards away. She turned on her tracks and started to make her way up the gully— but the stranger had already winded her kill and sensed her presence : in a flash he had leaped down the side and cut off her retreat.

Now she was ready to give battle, for no tiger could drive her away from her cave without a struggle. She licked her great jaws and crept cautiously on to an overhanging rock, so that when the attack came she should be above and not below her enemy. For a moment she lay still, watching the shape that stood out clearly on the track in the moonlight. It was staring at her now, motionless. She decided to spring first, and gripped tight to the earth. Before she could take off into the air, the tiger seemed to sense her plan, and growled softly.

There was no invitation to battle in that growl. The tigress stood erect for a moment and did not move until the tiger had growled again. Then she dropped lightly to the track and took a few paces towards him—before her stood her mate.

The tigress bared her teeth and tossed her head in the air. She uttered a low growl, but her eyes were gentle. ' My lord, I have killed a fine buck—go, satisfy your hunger,' she said.

While the tiger snarled and growled over the remains of the kill, the tigress withdrew, for a tiger does not like to be watched by his womenfolk while he eats. She sat down on a rock and carefully removed all traces of her feed ; then she waited patiently until her mate had eaten, drunk, and finished his toilet in his turn.

The tiger started at once to make for the direction of the cave, but for a moment the tigress hung back. She looked all around her, listening intently, and the tiger snarled at the delay. He had had a heavy meal and wanted to sleep it off.

' What are you waiting for ? It is time we got back to the cave. It is already beginning to get light,' he told her crossly.

The tigress took no notice of him, but stood gazing into the jungle in the opposite direction, her tail swinging from side to side. She wanted to leave the neighbourhood where she had

suffered so much, but the knowledge that their larder was full—
for the tiger had hidden the remains of the buck in a safe place
—persuaded her to return to the cave at any rate for a short
time.   After her mate had growled again she followed slowly in
his wake.

CHAPTER IV

Flight

ON THE following night they did not venture from the cave until the moon had risen. When they had reached the lower part of the jungle the tigress felt uneasy, and she became more and more nervous as they approached the cache where the remains of the kill lay hidden.

' Man ! Man has been here ! ' she snarled, her eyes gleaming with hate and fear.

The tiger stopped and sniffed the air. He too recognized the taint and paused in the track, lashing his tail. For a moment he seemed undecided, but suddenly he moved forward at a pace that made the silence of his passage little short of a miracle.

The tigress followed him reluctantly. She knew only too well what was to be expected from man, but she preferred following her mate to being alone again.

They reached the thick bush in which the carcase had been hidden, but both of them halted instinctively. The buck had been moved, dragged by men's hands, and laid just outside the bush in the moonlight. Each of them could see it as they peered through the tangle of the thicket.

The tigress glanced at her mate. ' I told you so ! ' her look said, but she made no sound. Then she lifted her nose in the direction of a bush that certainly had not been growing in that spot the night before. ' Here we are ! ' said that smell, as clearly as if the information had been shouted into a megaphone.

The two animals turned instantaneously, as if they were worked by a single piece of mechanism—there was a crash as they leapt into the jungle—then silence !

' That was the tigress ! ' murmured the shikari in a whisper. ' I told the Presence that she would return to the kill.'

The Maharajah, who had sat for hours in a *machan* with his finger on the trigger of his grand new English sporting rifle, was so startled that he fired off both barrels at once, while the weapon lay across his knees. The recoil knocked both him and the shikari backwards into a thorn-bush, severely bruising and scratching them both. The Maharajah scrambled to his feet. He was terrified out of his life, for he was certain that the tiger had landed on top of him. Throwing his gun from him he fled for his life down the track towards the village, while far away in the jungle came the mocking roar of the tiger, as he broke the neck of one of the village cows that had strayed from the herd near the watering-hole.

They ate their fill of the cow which had so fortunately crossed their path, and left the remains for a pair of jackals, who had been hiding in the bushes close at hand, ready to steal the kill as soon as the tigers had departed.

When he had drunk the tiger wanted to return to the cave, but the tigress had been nervous all that evening, and she lay down and growled angrily. ' There is danger here,' she repeated. ' I have been in the hands of men and I *know*. They have fled to-night—but who knows what will happen to-morrow ? They will return—as many of them as there are flies in the jungle—armed with all manner of things which I do not understand, and they will shut us up in little boxes and feed us on fouled meat. Phaugh ! I can remember that smell even now ! '

' They would never catch me ! ' boasted the tiger, with a deep-throated roar of fury at the very idea. ' I would kill them by one blow from my paw ! '

The tigress looked at him with her lip curling. ' Talk ! talk ! All talk ! ' she sneered. ' Man never lets you get close to him. All the time they kept me shut in that foul-smelling place they never came close enough to me—until one day—' she swished her tail and snarled, showing her huge pointed teeth. ' Even then I did not kill many of them,' she added. ' Go back to the cave if you wish—but go alone.' With another swish of her tail she dived into the forest and made her way at a slow trot into the heart of the jungle.

The tiger knew better than to argue with his mate when her mind was made up. He pretended that he was not frightened by her warning, but he knew instinctively that man was on their tracks and that their lair in the ravine was no longer safe. He paused for a minute or two longer, and then followed his mate into the jungle.

In spite of her heavy meal the tigress pressed on for mile after mile, until they had reached a thick belt of trees and scrub. Here she stopped, sniffing the air anxiously to make sure that there was no trace of the hated scent of man.

The tiger pushed her aside and took the lead. The moon had waned, and once more the jungle was plunged in darkness, but it made no difference to the speed with which they forged ahead. They pressed on until the sky had become tinged with the grey streaks of light which herald the approach of dawn.

They had now left the forest behind them, and were crossing a wide plain covered with thick, tall grass. Presently the sun peeped over the horizon, and the steel sky seemed to blush a rosy pink. Across the plain, smoke from a large village was drifting, like the cold-weather mists that linger over the open country until the heat of the sun has dispersed them.

A pair of crows, sleeping in a thorn bush, woke as the first tinge of dawn crept into the sky. Their beady, boot-button eyes marked a ripple in the grass. ' What is this ? What is this ? We must find it out immediately ! ' cawed one of them.

A crow can never mind his own business, and even when he has satisfied his own curiosity he broadcasts his information to the rest of the jungle.

The crows shook the sleep out of their feathers, stretched their wings and rose into the air. It did not take them long to find the cause of that ripple in the grass, and in a few seconds they were joined by all their friends and relations in the neighbourhood. ' Tigers ! Tigers ! ' shouted the crows shrilly, wheeling above the heads of the infuriated beasts. In vain the tigers quickened their pace, trying to outstrip their tormentors : in vain they reared up on their hind legs and snatched wildly at the little black bodies that hovered just out of their reach.

' We must get back to the jungle,' snarled the tiger, as he missed a crow by inches. ' May they die of mange and be devoured by ants ! May their feathers fall out and their bodies be torn in pieces by monkeys ! '

' We cannot go back—we *must* go on ! ' growled the tigress through set teeth. ' There will be jungle beyond this plain ! Take no notice of them ! '

It was broad daylight before the tigers reached the shelter of some thick undergrowth and were able to shake off their noisy escort. Thankfully they buried themselves in the cool green shade of the jungle, caring nothing for the hysterical chatter of the monkeys in the trees. They knew that the monkeys, unlike the crows, would soon get bored and swing away through the branches to find some other kind of amusement.

The sun, even at its hottest, could scarcely penetrate the thick tangle of creepers where the tigers had found shelter, so that they were completely invisible—nevertheless they slept fitfully, for they might be trespassing on some other tiger's hunting-ground, and in that case it would mean a fight to the death if they were discovered. At every little sound both tigers opened their eyes and pricked their ears—but each time it was a false alarm.

For a whole day they lay in hiding, and, as soon as the sun had set, the tigress got to her feet and nosed her way out of the tangled creeper, with the tiger at her heels. She stood outside their shelter, her nose tasting the wind and her tail slowly swinging.

' Why so much hurry ? ' asked the tiger, as he reared himself against a tree and started to sharpen his claws. ' Why not stay here ? I am ready to fight if necessary.'

' No ! no ! We must go on—go on ! ' growled his mate.

The tiger dropped to the ground again, but in spite of his enormous size he did not make a sound. ' You are hard to please,' he grumbled. ' Let us be going.'

Their first care was to obtain a meal, but game seemed to be scarce in that part of the jungle. Mile after mile they covered, slipping through the dark shadows of the trees, unheard and unseen—but still there was no prospect of a kill.

' Is there no game in this haunt of monkeys ? ' growled the tiger. ' I am hungry.'

The tigress did not answer him. She had dropped flat on her belly, and the tiger instantly did the same. Both animals listened intently : a faint wail was coming from a clump of trees that loomed up at the side of a clearing. The tiger's lips twitched back in a grin of anticipation. ' Goat ! ' he murmured.

They rose to their feet and moved towards that miserable sound. Suddenly the tigress stiffened. ' Man ! ' she growled—so low that only the tiger heard her.

Her mate was too hungry to heed any such warning. With infinite caution he continued to move forward and the tigress followed reluctantly a few paces behind him. As the goat, which had been tethered to a post in a small clearing, scented the tigers it became nearly frantic with terror.

' The poor old billy has winded something ! ' whispered one of two men hidden in a *machan* high up in a neighbouring tree. ' If only that infernal moon would come out we might get a shot.'

For a minute the men waited, listening to the bleat of the wretched animal, until there was a silence so sudden that it seemed as if the beast had been pole-axed.

' He's got him ! ' said one of the men. He fired blindly in the direction of the spot where the goat had been tethered. There was a rustle in the undergrowth, followed once more by deep silence.

CHAPTER V

Sanctuary

IT TURNED out to be a very old and bony goat !

The tiger had broken its neck with one blow of his paw, and had snatched at the carcase so greedily that he had torn out from the earth the peg to which it had been tethered. The flash of the gun, and the echoing report, terrified him, although the bullet had hit the ground yards away. He bounded to one side, and made off into the jungle, dragging the goat with the rope and peg attached.

Very soon the tigress joined him, and together they plunged on until they considered that they had left the clearing so far behind that they were perfectly safe. Here the tiger dropped the goat, and turned to the tigress with a growl.

' This is my kill ! Go to one side until I have satisfied my hunger ! ' he said.

But the tigress, who, had the kill been of a respectable size, might have obeyed her lord's command, had no intention of leaving him alone with the goat. She felt that there would be nothing left for her but the rope and peg if she did not share in the meal at the beginning.

' No ! I will feed *now* ! ' she snarled defiantly.

For several minutes the two animals faced each other above the dead goat, their teeth bared and their eyes savage with rage and hunger. The tiger bent down first and tore a huge piece of meat from the goat's haunch. As soon as he had raised his head the tigress in her turn tore off a hunk and dragged it back a little way, where she devoured it in a few seconds.

It was not a pleasant meal, and even when there was nothing left of the goat but its horns neither of them had satisfied their hunger. But as soon as there was nothing left to fight about they forgot their quarrel, cleaned themselves carefully, and set out again on their wanderings.

Mile after mile slipped by until they had reached a block of jungle that neither of them had been in before. For the first time since they had left their kill the tiger spoke.

' There is game here—game in plenty,' he growled, sniffing the air.

' Yes,' answered the tigress, lowering her head, ' but it is the hunting-ground of another ! You must fight him for it. Here it is that I want to live.'

They moved forward again, slinking along with the greatest caution, until the tigress stopped suddenly. ' Blood ! ' she said briefly. ' He has killed ! '

There was no mistaking that smell !

They made a wide detour and approached up wind the place where the strange tiger was resting after a feast. He was unaware of their presence until they were close on him, when he rose to his feet with a roar, and turned to face the trespassers in his domain.

He was a young tiger of about three years old who had only lately started hunting on his own, but, with the pride of youth, he felt himself capable of taking on any beast in the jungle.

His tactics were simple, for he rushed at the newcomer with a roar of fury, and both animals rose on their hind legs with the force of their impact. They fought like huge cats, tearing at each other with their teeth and claws, while the soft grass under their hind legs became ploughed with deep furrows as they dug in their back claws to get a firmer hold. Every now and then they fell apart and lay panting on the ground, exhausted by their efforts. The young tiger had youth and agility on his side, but he could not match the great weight and experience of the older tiger, who had survived many a fight and was still in his prime.

The young tiger backed a little and came rushing in once more. As he reared up the older tiger seemed to slip and rolled over on his back.

' The victory is mine ! ' thought the young tiger exultingly. ' Now I will tear him to pieces ! ' and he made to leap upon his prostrate enemy.

He realized his mistake too late. In an instant two huge forelegs were round him, huge sharp teeth were buried deep in his throat. The old tiger grimly disembowelled his enemy with the claws of his hind legs as a cat will tear out the heart from a ball of paper.

When the old tiger had pulled himself clear of the fight the late owner of the coveted piece of jungle lay like a bloody rag at his feet. ' Fool ! ' he muttered. ' Didn't your mother teach you that trick when you were a cub ? '

He moved away from the carcase and lay down in the shadow of a bush to lick his wounds. Long, angry red scratches and bites had scored his shoulders, and a deep gash ran from his right ear downwards towards his eye.

When he had licked his wounds clean of any poison that might have been gathered from his victim's claws he rolled in the mud so as to cover them against the attack of flies. Then he looked round for the tigress, calling to her with deep-throated growls.

Her answer came almost immediately. ' Here I am, oh mighty fighter ! ' she roared, her eyes gleaming with pride as she sniffed the body of his victim. ' I have found a perfect lair in which you may rest—'

' Is there water ? ' growled the tiger, for his wounds were painful and had made him thirsty.

' Only a few paces from where you now lie.'

The tiger rose painfully to his feet. He was stiff and tired ; but for his raging thirst he would have slept where he lay.

The tigress led him to the brink of a little stream, where he buried his jowl deep in the cooling water. At last he raised his head.

' Where is the lair ?　Lead me to it ! ' he ordered.

She turned meekly at his bidding and led him to a spot where a huge tree, hollow with age, had fallen across some boulders, forming a deep, cool cave.　The two animals crept into the hollow trunk.　There was no scent there, neither of beast nor man : they lay down gladly, to sleep through the hours of daylight.

CHAPTER VI

The God of the Jungle

THE TIGERS had unwittingly discovered a veritable paradise.
The forest abounded in game, their lair was dry and shaded from
the heat of the sun, and, as had been the case with their old
home, water ran close to their door. Furthermore, their nearest
human neighbour was a man whose very last wish was to kill any
of the jungle folk, from the largest to the smallest.

Mr Jeremy Mainwaring's bungalow lay in a large clearing
in the jungle. Round it was a garden which, in spite of the care
lavished upon it, never did very well—the beasts in the forest surrounding it had a great liking for tasty young shoots and, as the
gardener pointed out with bitterness to his friends, the sahib
would do nothing to stop them.

Mr Mainwaring was Forest Officer ; his job was to super-
vise the planting of new areas and the felling of timber ; his
hobby was the study of the wild animals to whom the forests
were home.

Certainly he was sorry when all his young pea-shoots were
stolen by the little picking hands of the monkeys, but he felt
amply rewarded if he could get a good photograph of them doing

so. Dainty spotted deer would leap over the low wall of the garden, tempted out of their usual caution by the delicious things that grew beyond. If it had not been for the gardener, or *mali*, who had been known to sit up all night guarding a melon, nothing would have survived in the garden for a week.

The *mali* was the father of a large family, and he would hide his reluctant children about the garden during the day with orders to shoo away any beast that tried to sample the plants. At night, unknown to his master, he would tie tins full of stones, that rattled at the least touch, to poles set in rows beside the plants which he wanted to protect. These were uprooted first thing in the morning and carefully concealed in his hut.

But though he waged continuous war against the beasts of the jungle he most hated those chosen animals who, though at liberty to go and come as they pleased, lived in or near the compound. They could hardly be called pets, for no kind of restraint was placed upon them, but the way in which they regarded the bungalow as their own was, in the opinion of all the servants as well as the unfortunate *mali*, outrageous. For some weeks a bear had taken up his quarters in one of the outside bathrooms ! He had been attracted in the first place by an ant's nest, which he had demolished in record time. The sahib made no attempt to turn him out until he was ready to go of his own accord. Even the leopard, most treacherous of beasts, who kills for the love of killing, was allowed to stalk his prey within a mile of the neighbouring village without any attempt being made to secure his skin.

Birds of all kinds nested in the eaves of the bungalow, monkeys did physical jerks through the arched verandah, and stole the food as the cook carried it from the kitchen to the house —and the sahib laughed !

Yet every man in the country for miles around considered it a privilege to serve Mr Mainwaring ; in fact many of them believed that he was a god—the god of the jungle—for they came from primitive tribes who inhabit the jungle that stretches up to the foothills of the mountains. They could find no other explanation for the fascination which he exercised over all animals. Who but a god of the jungle could go alone and unarmed into the forest, set up his camera in the very rocks used by the tigers, and come back safe every time ? Nothing touched him, neither the animals nor the spirits that roamed among the great trees after dark !

This belief was also held by the gangs of coolies who worked under Mr Mainwaring, with the result that he got more work out of them than any other man could have done. But Mr Mainwaring himself was completely ignorant of the veneration in which he was held.

One evening, a fortnight after the tigers had taken up their residence in their new home, Mr Mainwaring left his bungalow with his camera and flash-light apparatus under his arm.

The servants followed him with their eyes as he strode down the garden drive, crossed the wide clearing beyond and finally vanished among the trees.

As he disappeared the bearer shook his head. ' It is the young tiger whose picture he wishes to take ! Supposing the tiger does not know who he is ! I wish he would not go thus unprotected ! '

' Unprotected ! What nonsense ! ' scoffed the *khansama.* ' Unprotected out there among his own people ! '

Mr Mainwaring, unaware of the fear for his safety that stirred in the bearer's breast, walked up a narrow jungle-path. He knew the way well, but even so he did not omit any of his usual precautions. His ears, sharpened by his life-long acquaintance with the jungle and its noises, listened to every tiny sound.

The sun had not yet set, but under the trees the light was poor, so he trusted to his ears rather than to his eyes—he had found from experience that they were more reliable. A herd of elephants, grazing among the saplings near a thick belt of trees, carried on undisturbed, for they knew they had nothing to fear from the white man.

At last Mr Mainwaring reached the place he was aiming for—a narrow track, obviously of recent date, which cut across the old track on which he was walking. He turned up this new track for a short distance and then laid his burden on the ground. First he put the camera into position in a bush, so that the lens faced the path in the opposite direction to that which he had just come. Then he arranged the flashlight, and laid a tiny trip-wire across the path. He stepped back and surveyed his work. The least touch of the wire would now fire the magnesium flash and the camera simultaneously, and whatever beast touched that wire would be straight in front of the lens—and as Mr Main-waring earnestly hoped, facing it.

' Perhaps I shall get a decent picture this time,' he said to himself. ' I do hope it is the tiger again—the last one I got of him was not good, and the time before he was going the wrong way, and I only got a photograph of his tail.'

He glanced round carefully to see if he had left any avoidable evidence of his presence, and then turned and made his way back to the bungalow.

As he stepped out on to the clearing the sun set. For a few moments the sky was a sheet of flaming gold, spreading above the black mass of the trees ; then it disappeared. The night had come.

Mr Mainwaring turned and looked back into the forest. ' Choose that path again to-night, old chap,' he murmured to the darkness. ' It's three weeks and more since you came that way ! '

But only a few whitened bones were left of the tiger whom he was addressing.

CHAPTER VII

A Pilgrimage

As MR MAINWARING got into bed that night his last thoughts were of the tiger which had died so gallantly a fortnight before. Meanwhile his conquerors were eating their fill of a large buck.

The tigress had finished her meal and had made her way back to the lair, after summoning her mate to the feast. He, in his turn, ate as much as he could hold and then carefully hid the remains. Satisfied that his larder was safe from discovery he sat down to clean himself. The wounds that he had received in his battle with the young tiger were now nearly healed, but he had gladly left to the tigress the work of tracking and killing the buck. Ever since the fight she had hunted steadily so that he could regain his strength as soon as possible, and he looked with satisfaction in the direction of the cache—there would be no need for either of them to hunt for several days to come.

At last, every hair on his gleaming skin clean and shining, he rose to his feet, stretched himself and turned his head towards home.

Just as he mounted the slight incline which led to the lair, he paused. The urge to roam the jungle was on him and he half turned. ' I'll wait another night,' he thought. ' One more day's rest will be wiser—'

Two great bounds brought him to the mouth of the lair, but he could not enter. The tigress stood barring his way, her lips drawn back in a snarl. The tiger was so surprised that he stood stock still and looked at her in astonishment.

' Get out ! ' snarled the tigress. ' Get out and take some exercise, you lazy great brute ! '

' What on earth is the matter with you ? That is not the way to speak to me ! ' roared the tiger.

' You heard what I said,' the tigress roared back. ' Now go ! '

A string of monkeys sleeping in a tree not far off woke at the sound and chattered nervously.

' There's a nasty temper for you ! Did you ever hear any-thing like it ? ' said one, nestling closer to his neighbour for comfort.

' Now don't you go pushing me off ! I'm not afraid of tigers ! '

' Oh ho ! Listen to him ! Not afraid of tigers ! ' shouted the others all together. ' We are none of us afraid of tigers—yellow striped earthbound cats ! '

Again the tiger's roar shook the jungle, and the monkeys fell silent on the instant, shivering and huddling together in a paroxysm of fright. They were safe enough in their tree, as they knew perfectly well, but—and the ' but ' was in all their minds—

many a monkey has found its way into a tiger's tummy ! The thought was not a comforting one.

But the tigers were not interested in the monkeys. The tigress was in a filthy temper and her mate had already thought of wandering abroad again. He had far more self-respect than to stand arguing with his mate outside their lair ; he turned, bounded down the slope again, and disappeared into the forest.

He was glad that the decision had been forced upon him, for once on the march he found that the lazy feeling which had suggested another day's rest had vanished. It was long since he had been completely alone, and to his joy he found that his recently healed wounds gave him no trouble at all.

He saw and winded plenty of game, but he was not to be tempted to kill, for he had already gorged himself full, and a tiger does not hunt for amusement—only for food. So that while his passing through the forest that night caused a moment's panic in the hearts of the deer, none of them had any real cause for fright.

The tiger went where his fancy led him : sometimes following the game tracks : sometimes slipping through untrodden jungle : but always cautious, always on the look-out for a possible enemy that might be lying in wait by a tree, or hidden near the track.

Suddenly, as he walked on a narrow game track, something moved ever so slightly beneath his paw. He looked up, and on that instant there was a dull ' boom ' and a blinding flash—then the darkness descended again. The tiger waited for the roar of thunder which usually followed such a flash of light, but the jungle was still wrapped in silence.

He did not wait to investigate, but swung off the path in an instant and continued his journey through the undergrowth. Had he been alarmed by the flash he would have sprung in its direction, but he was not. For a few minutes a sense of bewilderment lingered, and then all recollection of it passed from his mind.

When Mr Mainwaring went to visit his trap early the next morning he found to his joy that it had been fired. He examined the ground along the path, but the weather was dry and there was no sign of any tracks. He picked up his camera and went home as quickly as he could.

Mr Mainwaring had had too many failures to be wildly anxious as to the result of his night's work, but he still retained the power to be thrilled by success, even if he had acquired a patience as great as that of the animals he studied. As soon, however, as the first trace of a pattern would appear on the face of the plate as he slowly rocked the tray of the developing solution in his darkroom, he would become as enthusiastic as a schoolboy collector who sights a strange butterfly, or discovers a rare postage stamp. Even the most experienced tiger could not have displayed more patience that day, for he locked the camera up in a cupboard and went about his day's work as calmly as if such things as darkrooms and developing dishes did not exist.

It was not until the following afternoon that the printed result lay in his hands. As Mr Mainwaring looked at the photograph he could not help a gasp of admiration. ' I shall never do better than this ! ' he said aloud, as he gazed at the picture.

The great eyes of the tiger looked up at him with an expression of wonder, his ears pricked, his body taut, ready for a spring if a spring should be necessary.

' Beautiful ! Beyond words, beautiful ! ' murmured the forest officer. ' But, my lad, you are not the tiger I was expecting ! '

CHAPTER VIII

The Coming of Argh

MEANWHILE THE subject of the photograph continued his travels, with no idea of the sensation that that curious flash of lightning had caused. For three weeks he roamed the country, lying up during the hours of daylight in any place that took his fancy, but always working round in a wide circle ; so that when his wanderings were over he found himself back at the lair.

He bounded up the slope, to be brought up suddenly by the sight which met his eyes at the top.

Outside the lair lay the tigress, stretched full length in the light of the moon, while round her tumbled two little furry bundles. As the tiger appeared over the rise the tigress called the cubs to her with a deep-throated growl, which they obeyed immediately.

' It is I,' announced the tiger, stepping into the moonlight. ' Am I welcome ? '

' You are welcome ! ' answered the tigress. ' These are our cubs. Look them over ! '

The tiger stepped across the intervening patch of grass and examined each cub in turn with the greatest care. The cubs, frightened by the size and strangeness of their father, mewed piteously and tried to run to their mother for protection ; but she would have none of them. With a low growl she shoved them back towards the tiger, who, to their surprise and relief, did no more than roll them over in exactly the same way that their mother did, and smell them.

At last he was satisfied. ' They are fine cubs,' he said, and lay down at a little distance from his mate.

Now that the examination was over and their mother no longer drove them from her, the cubs huddled close together and examined their father in their turn with wondering, apprehensive eyes.

He was certainly worth looking at. His huge head, fringed with a magnificent ruff, lay on his paws, while behind rose the massive bulk of his powerful shoulders. Even in his present peaceful attitude his enormous strength was apparent to the most inexperienced eye, and the cubs gazed at him and marvelled. He seemed even to dwarf their mother !

Their eyes grew rounder and rounder as they listened to the deep rumbling voice with which he talked to his mate, and saw his enormous white teeth when he yawned—which he did frequently. Neither of the pair took the slightest notice of their cubs, but the tiger was well aware of their fascinated scrutiny— and was not a little pleased by it.

' You did not meet—man ? ' asked the tigress.

The tiger rolled over on his side and yawned luxuriously. ' I met no man—I fear no man ! ' he answered.

The tigress's eyes darkened. ' You have not had the cause to fear—man—as I have,' she murmured. ' What of the game ? '

' Plentiful—plentiful. Enough for our needs for many moons to come.'

' That is good,' said the tigress. ' Tell me now of the game runs—'

The two tigers talked on, and bit by bit the cubs' first alarm subsided. Their natural energy would not allow them to sit huddled against their mother any longer. Cautiously they left her side and started to play in the moonlight, but all the time with a wary eye on the huge stranger who had arrived so unaccountably in their midst. He had a fascination for them that they could not resist. Gradually they edged nearer and nearer : first a little drunken wobbly run towards him, and then a precipitate retreat, rolling over each other in their anxiety to reach safety. But as the great tiger did nothing they became bolder.

One of the cubs was much larger and heavier than the other, and also more venturesome. As he would one day be a real tiger himself he felt that he must show his little sister how a tiger tackles a problem. Summoning all the courage in his little body he advanced upon his mighty parent—closer and closer he came, his sister following apprehensively a little way behind him. Soon he was within touching distance of his father's tail. He sat down and eyed it with interest. It was twitching in the most engaging manner—what would happen if he gave it just *one* little pat ?

He cocked his head on one side and glanced at his sister out of the corner of his eye. She had sat down out of reach and was watching him with round admiring eyes, ready to fly at the least sign of danger. Emboldened by that look he stretched out a paw and patted the tip of his parent's tail—and nothing happened.

' See ! Aren't I brave ? ' said the look he gave his more timid sister.

The tigress, who had been lying flat like her mate, raised herself on her shoulder.

' The cubs will be tumbling over your neck in a moment. *Remember* that they are your cubs if they should nip you where it hurts,' she warned him.

It was as well that he had been warned, for finding that nothing happened, the cubs got bolder and bolder. The big cub hauled himself up on his father's back, using his baby claws without the slightest consideration. Actually the tiger scarcely felt it, but when his son slipped on his great neck and rolled down his face, holding on to his soft lips with little needle-like claws, it was a different matter. Quick as lightning he raised his huge paw to sweep the little nuisance off, but, quick as he was, the tigress was quicker. In a flash she had the cub by the scruff of his neck, and had whisked him away before his father's great paw, with every claw extended, smashed through the empty air.

The tigress dropped the cub on to the ground and gave it a gentle cuff over the ear. ' That is to teach you to be more careful, you silly little thing,' she said. Then she turned to the tiger.

'You must keep your temper better than that!' she snarled. 'You would have killed him if you had hit him!'

The tiger grumbled to himself, but made no answer. He was certainly sorry, but he was glad that his mate had smacked his tormentor's head. He was not so sure that he liked being a father.

CHAPTER IX

First Lessons

IT DID not take the cubs long to learn just how much of their attentions their father would stand. When he was full of good food, and therefore in a good temper, he was quite ready to play with them, but the cubs were very wary when, as sometimes happened, the tiger had not enjoyed his meal.

Their life in their new home was indeed a pleasant one. Food and water were plentiful, and since the district was far away from any road or railway no man had so far invaded what would have proved to be a sportsman's paradise.

Had the tigress but known it, a man was their greatest protection, for Mr Mainwaring was a past master of evasion. His wonderful photographs lay hidden in a locked box in his bungalow ; he had no wish to publish them and turn the glare of publicity on his jungle.

When he was questioned about the quantity of game in his part of the world he would put on an air of engaging frankness and say that he would welcome a party at any time, but he must warn them that never once, all the years he had been there, had he shot a thing worth shooting. This statement was strictly true, although misleading : when he killed it was only as a matter of execution. He was a magnificent shot, and if he heard that any beast, either through old age or injury, was making himself a terror or a nuisance to the natives, he would set forth with his gun-bearer. There was no fuss and no excitement, and when they returned the skin of the offender came with them and peace reigned in the villages once more.

Beyond the reach of their greatest enemy the tigers lived their happy, peaceful lives. The cubs grew fast and soon their legs were strong enough to allow them to explore the forest round about the lair. They were never allowed to go far without their mother's protection, and if they strayed a deep, rumbling growl would bring them back as fast as their legs could carry them. A full-grown tiger has not much to fear, but a cub might easily fall a prey to a snake or a prowling leopard.

One beautiful morning the tiger cub strolled down the slope below the lair. The jungle was full of exciting noises which stirred his blood in a way he had not experienced before. He glanced back over his shoulder at his mother. No ! She was not looking at him.

He bounced down the slope, and crossed the greensward where his father had fought for, and won, their present home. A large leaf fluttered down from a sal tree, and the cub pounced on it, throwing it up over his head and catching it again. For a few moments he played like a kitten, and then, tiring of such childish amusement, abandoned the leaf and continued on his voyage of exploration.

A rustle in a bush attracted his attention and he stalked forward as his mother had taught him to do. His steps were wary, but his brain was bubbling with excitement. ' This is what I came for ! I am a mighty hunter—now I will make my kill ! '

Inch by inch he crept on. He could see the brilliant plumage of a jungle-cock that was sitting on a low bough of a bush preening his feathers. The cub dropped flat to the ground, his head low and his eyes fixed on his quarry—another inch and he would spring ! Suddenly his mother's voice roared through the trees, ' Come back, my son ! '

There was a moment's silence before the order was repeated, this time more sharply.

The cub still took no notice, for at the first sound of his mother's voice the jungle-cock had whirred away and he, with head thrown back and lips quivering hungrily above his white pin-pointed teeth, was following the line of its flight with eager eyes. But the boom of his mother's third summons brought his thoughts back to earth with a jerk. She was angry now, although there was a faint tinge of alarm in the roar which echoed and re-echoed through the forest.

' Coming ! I am coming, mother ! '

With one last look in the direction which his prey had taken the cub turned and began to retrace his steps. But he did not get far before his mother came rushing towards him like a yellow and black rocket.

' Why did you not answer my first call ? ' she demanded furiously, when a glance had satisfied her that he was not hurt.

' I was hunting ! ' answered the cub, trying to get a grown-up growl into his voice, but only achieving an impertinent whine.

A blow from his mother's paw sent him staggering on to his haunches. ' Why should I not hunt ? Am I not a tiger ? ' he snarled peevishly.

' You are a cub—and cubs *obey*. Also tigers do not hunt for pleasure, but only for food. You have already fed—would you imitate the lowborn leopard and hunt for the sake of killing ? ' growled his mother. ' Go home ! The sun is up and it is time you were asleep.'

With another, though gentler, blow she started him on his way, following behind him in case he should turn aside. But the cub had no desire to defy her again. His side tingled with the blow from her paw and his feelings were hurt by her taunt, for no tiger, however young, likes to be compared to a leopard. To add to his humiliation, his father had watched the scene with eyes that glinted with amusement.

With one final shove the tigress pushed her son right to the back of the lair and lay down across the mouth.

The next night brought balm to his wounded feelings. As soon as it grew dusk the tiger had left the lair alone ; he moved silently along a track which led to the far side of his hunting-ground ; but the tigress did not leave for many hours afterwards.

The cubs fidgeted restlessly as they played outside. Was there to be no dinner to-night ? But at last the tigress rose and stretched herself.

' You shall accompany me to-night,' she said. ' The cache is empty and I must kill. You shall watch and see how it is done. Follow me ! '

She led the cubs through the jungle, stealing silently through the shadows. The cubs followed in her wake, copying every movement of her body. Presently they came to the edge of a wide plain, and the tigress suddenly dropped flat in the shadow of a bush. The cubs quietly followed her example, wondering what was going to happen next.

With deep interest their eyes followed hers and saw a fine Sambur standing knee deep in the long grass ; a little beyond him stood the rest of the herd. The cubs watched, quivering with excitement. Each moment they expected to see their mother pounce upon him, but she made no move.

The buck's head was raised : perhaps he heard some faint noise or caught for an instant the scent of tiger. Still as a rock he stood, testing the air with his sensitive nostrils, but the morning breeze was blowing the scent of the tigers away from him, and he lowered his head and went on grazing.

Now the tigress began to ease herself forward. Inch by inch, her body pressed against the ground, she moved nearer and nearer to her prey. Another few seconds were spent in shifting her position until every muscle was correctly balanced ; then, with the swiftness of a bullet, she shot through the air on to the back of her victim and buried her teeth in his neck. He fell to the earth with a crash—dead in an instant.

For a moment the tigress lay beside him, her tail swishing and her eyes gleaming red, while the rest of the herd galloped to safety, for they had started off at the first sound of the roar she had given while she charged.

When the thunder of their hoofs had died away the tigress turned back to her cubs, who had watched with feverish excitement.

' Come ! ' she called.

They needed no second bidding, and with growls of joy sprang at the haunches of the dead buck.

For a moment she watched them. ' It is thus we kill in the open,' she said, ' but should you kill beneath trees leap at the throat, not the back, lest when he falls he kill you also by crushing you against a tree. Come, I will now drag him into cover.'

She seized her victim by the throat and pulled him through the grass, while the cubs bit and worried at his heels.

Once safely hidden in the jungle the tigers made their meal. With growls that were a faint imitation of their mother's the cubs dragged at the haunches of the kill. This was the real thing, so different from tearing the small portions of meat brought to them outside the lair.

' Soon, soon I shall be able to kill for myself ! ' boasted the tiger cub.   ' It is easy ! '

' It will be many moons before you can drag down a buck, my son,' answered his mother, as she licked her cheeks and whiskers.   ' But you shall soon start to kill small game ! '

With a grunt she seized her small son with one paw and started to give him a thorough washing.

CHAPTER X

Man

WHEN THEIR larder was full the tigress spent the nights teaching her children how to hunt for themselves. They learnt the art of lying as motionless as a log, watching their prey until the moment came when a spring was certain of success.

Birds and small animals were generally their victims. The tigress taught them how to kill the porcupine, that dainty beloved of tigers. They learnt to follow him with infinite caution ; to stalk round to his head and dispatch him with one blow on the nose, and so avoid the armoury of sharp-pointed weapons with which his back was covered.

Their mother was gentler now : the calm life that she had led since her escape from the rajah's palace had soothed her nerves, but the danger of man was ever present at the back of her mind. Her experience had been bitter, and she was determined to impart to her cubs as much of her knowledge of man as she could.

One morning, instead of taking them back to the lair, she led them through the jungle in a direction they had not been before. A langur, enjoying the early morning sun from his perch on the top of a fallen tree trunk, watched them depart ;

he felt too lazy to follow, although his curiosity nearly got the better of him.

The cubs were puzzled. ' Where are we going ? I want to go home ! ' whimpered the baby tigress. She was a lazy little person, and had eaten rather more than was comfortable. Her one idea was to get home and sleep.

The tigress took not the slightest notice of her, but the tiger cub, full of mischief, nipped his sister in the flank. The next minute they were rolling over and over, biting and scratching. This time the tigress did pay attention. She cuffed them apart, snarling angrily.

' Follow me *silently*, ' she commanded. ' You are now to see your greatest enemy—follow me, and make no sound ! '

' Will you kill the enemy ? I will help you ! ' cried the tiger cub, bouncing round his mother's heels.

' No ! ' growled the tigress.

The cubs, startled by the fury of her reply, relapsed into silence, and when their mother continued her walk through the jungle they trotted noiselessly at her heels.

On, on they went—the tigress grim, and silent. The sun had risen and soon its rays began to filter through the thick undergrowth through which they were marching. At last the tigress called a halt. Deep in a thick bamboo-clump she lay down, and motioned to the cubs to do the same. They looked at her with wonder, puzzled by the nervous tension of her manner and the curious gleam in her eyes.

Outside the clump, just discernible through the stems of the bamboo, ran a rough track, and it was on this track that the tigress's eyes were fixed. The cubs stared at the track as if they were gazing into a crystal. From time to time they blinked at each other, twitching their ears, filled with wonder as to what manner of animal it might be that could cause their powerful and courageous mother such apprehension.

' It will be very large and fierce,' thought the tiger cub to himself. ' Something about the size of a bull elephant, with the eyes of a snake and the temper of a leopard. But I am not afraid —no, I am not afraid ! '

He went on repeating the last words to himself, for he was very, very much afraid and only his complete faith in his mother kept him from rushing wildly away into the jungle.

At last ! A faint thud, thud, coming along the track towards their hiding-place ! He felt his mother stiffen, and saw the hair on her back rise in a bristly ridge.

' I am *not* frightened—*not* frightened,' he repeated again and again, his hair also bristling and his little eyes round with terror.

Nearer and nearer came the thudding noise, and then round the bend of the track came four figures. The hair on the tiger cub's back lay flat again, and the terror died out of his eyes. ' Four hairless monkeys ! The beast is yet to come,' he thought.

The coolies drew level with the bamboo clump and passed within a couple of yards of the watching tigers.

' Why, they haven't even got tails ! ' thought the cub ; then he turned his head and looked at his mother. She also was watching the hairless monkeys, but without her son's unconcern. Every hair on her body stood on end ; her eyes glowed like hot coals, and her lips twitched back from her teeth in a silent snarl. She watched them out of sight, but it was not until the sound of their footsteps had completely died away that she spoke.

' That was—MAN ! ' she growled. Then she rose slowly to her feet while her cubs looked at her with wondering eyes. What had come over her ? Was this their magnificent mother, trembling at the sight of four, dark brown, hairless, tailless animals padding through the dust on their hind legs ? But the look in her eyes forbade questions. They followed her in silence as she slunk out of the bamboo-clump and back into the jungle.

They were dead weary when they reached the lair, but their mother would not allow them to sleep.

' Now you have seen—Man,' she said, as she dropped to the ground outside the mouth of the cave.

' What of it ? ' asked the tiger cub cheekily. ' Even *I* could kill that thing ! '

' Peace, fool ! ' snarled his mother, with such savageness that the cub drew back, his ears flattened for the smack that he expected ; but no smack came.

' That is—Man,' she continued. ' He is more to be feared than any living thing—even the snakes that lie hidden in the grass : more also than an elephant cow whose calf has been killed. *He* kills for pleasure—*he* gives pain for pleasure. Alone he is harmless—but touch a man and all his tribe rise up in vengeance. I—I myself have been in the hands of men ! We have both, your father and I, been attacked by man with loud noises ! They bring the great fire, they scream, they smell— but they must not be touched. And they cannot be eaten ! '

The hate and fear in her voice sent a thrill of terror down the spines of the cubs. It was hard to believe that such a harmless-looking creature should be a danger to them, undisputed kings as they were of the jungle.

' Never forget my words—kill anything you have strength to kill—but never touch man ! '

' We will never touch man ! ' answered the cubs solemnly— then they yawned, and were fast asleep before their mother could say another word.

For a few moments the tigress watched them, and then she too dropped her head on her paws and fell asleep.

CHAPTER XI

Fire !

THE TIGER was the first to notice the approach of the hot weather. He became easily annoyed, and his young family soon learnt that it was unwise to tease him when he lay at full length in the shade, trying to keep cool. Soon the cubs too began to feel irritable, and their romps ended as often as not in a fight.

They could not understand the change that had come over their beautiful jungle. All the colours seemed to have disappeared : the air that they snuffed was hot : the delicate scents had gone : the stream that ran by their home dried up until it was little more than a few turgid pools linked together by a thin trickle of water ; and soon even that trickle of water vanished.

In the daytime they lay right at the back of the lair, with their panting bodies close to the rock. The atmosphere of the cave was fetid, but, if they poked their noses outside, the blast of hot air would beat in their faces like a flame. Only at sundown was there a light breeze, but once the sun had set it died, leaving the air even hotter than before ; so that they longed for daybreak during the night, and spent the day wearying for the evening breeze.

The only people who seemed to enjoy the heat were the monkeys. They played 'touch last' through the topmost branches of the trees, yelling at each other all the time. The family of tigers sweltering below them provided a constant topic of conversation, and the crows, their beaks agape with heat, joined in the monotonous chorus : ' Tigers ! Tigers ! They can't reach us here ! '

There was no privacy, no peace for the tigers throughout the broiling day except in their lair, for even if they managed to escape from the peeping eyes of the monkeys, the black beady eyes of the crows would seek them out. Several times they moved from one hiding-place to another, and now that they were strong enough the cubs accompanied their parents on these trips ; but, like the royalty they were, they could not escape the notice of the mob. Their every move was reported and commented upon.

Early one morning the tigress and her cubs were lying on the edge of the jungle, idly watching the red glow of approaching dawn as the sun rose above the horizon. Behind them black thunder-clouds lay banked thick and dark, spreading slowly over the sky : a sky full of beauty and terror.

The tigress was restless and uneasy, and communicated her agitation to the cubs, for on the air that blew across the grass plain, just beyond their retreat under a thick bamboo-clump, came a scent that filled her with foreboding—that scent that brings terror to the hearts of all the jungle folk from the weakest to the strongest—the smell of fire !

Presently a herd of deer shot past them, caring nothing for the tang of tiger that hung in the breeze, for they were flying from an even greater terror. The deer were followed by a sounder of pig, hurrying along, grunting with fear as they plunged among the trees. The tigress rose to her feet suddenly and, having sniffed the air again, set off in the wake of the pig, with the cubs following at her heels.

' What is it ?   What is it ? ' whimpered the tiger cub, as he stumbled along at his mother's heels.

' I am frightened, I am so frightened ! ' wailed the baby tigress piteously.

The tigress turned at the sound of her cubs' distress and urged them with low growls to hurry after her. But they needed no urging—they were hurrying as fast as their young legs would carry them—tripping over creepers, falling over the roots of trees, unable in their blind panic to look where they were going.

There was no wind to carry the drifting smoke towards them, nor were there sparks of charred wood and burning grass to warn them of their peril. The fire descended upon them with a sudden roar, as if it had been conjured into being by a magician. The tall grass was blasted away in a flash, leaving a blackened waste of smouldering earth.

The trees at the edge of the jungle crumpled in the face of that terrible sheet of flame : one moment the leaves hung lifeless and dry, the next minute a fountain of fire seemed to shoot up the roots to the smallest leaf at the top. It died down as quickly as it had shot into the air, and, leaving the blackened wreck behind, passed on to the next tree.

In a few minutes the fire had reached the thick jungle, and now a raging furnace blazed in the wake of the fleeing dwellers of the forest.

The air was thick with a black, oily smoke that stupefied beast and bird ; the monkeys climbed in terror to the topmost branches, only to be overcome by the smoke and to drop lifeless into the flames below ; above the trees the birds fluttered in terror, their tiny brains addled by the acrid fumes, until, as though drawn by an invisible hand, they too vanished among the tongues of flame ; the fire licked up greedily every living thing that it could reach.

Ahead of the fire marched the jungle—a mighty, ill-assorted army drawn together by the common panic ! Tiger and leopard raced side by side with pig and deer, while at their feet snakes of all sizes slithered over the ground like quicksilver ! Every eye was glazed with fear, but love of life drove them onward—away from that terror that was laying waste their home, no matter where the trail might lead.

High above the forest the vultures hovered, waiting their chance. Rain was coming, and with the rain they would reap their harvest. Many a feast was lying ready for them as soon as the fire had died down, and already the sky was being split by jagged forks of lightning, and shook to the deep rumble of thunder. Soon the storm would break in all its fury—but until then the vultures, invisible black specks against the inky sky, would wait, riding the air like thistledown, their eyes watching every movement in the terrible scene below.

The tigress was gradually being left behind by the rest of the animals. In spite of her terror she would not leave her cubs. Once, when a branch, flaming like a torch, came hurtling through the air and landed close to her, singeing her as it fell, she gave way to her panic, and deserted her cubs in a frenzied race for safety—but she did not go far. As she bounded away the cubs set up a wail of misery, and after a few yards she hesitated, turned, and retraced her steps.

' Come, come quickly ! ' she screamed frantically, but the cubs were now dead beat. The little tigress sat down and whimpered piteously, leaning against her brother for support. The tigress looked around her. Little trickles of flame were running along the ground, the forerunners of a mighty wall of flame, and soon the trees above them would burst into flame. Smoke filled her eyes and terror reigned in her heart, but even terror could not crush her love for her cubs. She seized the baby tigress by the scruff of her neck and, with the courage born of despair, set off again. The tiger cub tottered along weakly at her heels.

After she had carried the baby tigress a short way she returned and tried to do the same for her son, but he was too heavy for her in her wearied, half-suffocated condition.

In her despair she got behind the cub and pushed him along with her head, nipping his legs as she did so, and forced him along to the spot where she had left his sister—but it was slow work. Twice, three times she repeated the manoeuvre, but at last she was forced to give it up.

Her eyes, red-rimmed and watering, saw only the little ash-strewn body of her son lying where he had fallen ; behind him raged the white flames of the burning jungle : she was blind and deaf to ought else—blind to the jagged lightning cutting the black sky like a fiery ribbon : deaf to the peals of thunder. She heard only the roar of the fire and the crackle of the bamboos as they burst in the heat. With an agonized glance at her son for the last time, she picked up the smaller cub once more and staggered on her way.

On, on she went. Every step was an agony, for the soft pads of her feet were badly scorched. She was now oblivious of everything : she did not even notice when the black clouds above her burst at last, and the rain came hissing down in a solid wall of water ; and it was not until her weary legs collapsed beneath her that she sank to the ground.

For long hours the tigress and her female cub lay on the rain-soaked earth, past caring what fate had in store for them. When the storm had driven past them the sun shone on them once more : the hot rays revived the tigress, drawing up the moisture from her steaming hide ; she raised her head and gazed fearfully behind her.   There was no sign of that terrible glare ; there was no trace of smoke to choke her—and there was no trace of her tiger cub, though the female was safely curled up by her side.

Painfully she dragged herself to a sitting position and started to lick herself.

CHAPTER XII

Rescue

THE BABY tiger had not noticed his mother leave him : he lay in a heap, choked by that terrible smoke, too exhausted to move. If the fire had caught him then, he would have died without feeling the sting of the licking flames. Even when the torrent of rain broke loose over the jungle and lashed his little body until it quivered, he did not feel it. He did not hear the hiss with which the rain extinguished the fire, nor the crash of the thunder as clap followed clap with barely a pause between each.

For an hour the storm raged. Again and again the lightning struck the forest. Sometimes a tree, untouched by the fire, would topple headlong—the noise of its fall drowned by the roar of the thunder. Sometimes the blackened trunk would remain standing, its leafless branches stretching out uncouth limbs to the sky.

Throughout that inferno of storm the cub lay as though dead. The storm passed, the black clouds rolled back like a curtain, and the sun smiled benignly from a sky of purest blue, but still he did not move.

Then from that peaceful sky the vultures began to drop like black stones : there was many a sad, charred little body lying in the blackened wreck of the forest that the fire had not had time to consume, and the hideous birds clamoured to finish the work that the fire had begun.

Mr Mainwaring had ridden off to the threatened district as soon as the news of the outbreak had been brought to him. Several little clusters of huts lay in the path of the fire, and the owners had to be warned. From early dawn he laboured, riding from one village to another, here warning the occupants to be ready to fly if necessary, and there quelling the beginnings of panic. With anxious eyes he watched the gathering thunder clouds : if they did but break in time the villages would be safe, but, light though the breeze was, it was blowing the fire directly towards these villages. Even his own bungalow lay directly in its path and could not escape ruin.

A little group of villagers crowded round him, anxious for their homes and their crops.

' Sahib ! Sahib ! will the rain come in time ? ' they implored, their voices vibrant, verging on panic.

The forest-officer looked at the fire glowing against the black sky, and then at the sky itself. ' Yes, it will come in time,' he replied, inwardly praying that his judgment should prove correct.

His hearers hailed his answer with a shout and rushed back to the villages. ' The jungle god has spoken, our homes are safe ! ' they announced, and promptly sat down to prepare their morning meal as though there was nothing unusual occurring.

The suddenness with which the storm broke and put out the fire had taken Mr Mainwaring by surprise, and by the time he had reached the shelter of his bungalow he was soaked to the skin. ' In time ! ' he murmured to himself, as his bearer removed his wet clothes. ' In time—and even heavier than I had dared to hope ! '

As soon as the storm was over he went out again to estimate the extent of the damage. He went on foot, his legs encased in long boots, for he knew that the ground would be unsafe. Snakes, driven first by fire and then by rain, would be seeking sanctuary, and woe betide the man or beast who had the misfortune to step on one of them.

Mr Mainwaring looked sadly at the charred, evil-smelling remains of what had so lately been a forest of living green. He watched the vultures dropping from the sky—vile undertakers, ready to do their work as soon as their services were required.

It was a group of vultures that drew his attention to the tiger cub. He had been partly revived by the sun and was snarling weakly but furiously at the great birds sitting around him patiently in a ring, waiting for him to die. They barely moved aside when Mr Mainwaring strode into the circle and bent over the cub.

' Poor little chap ! ' he exclaimed, as he gazed at the singed, soaked little body. ' You have had a bad time of it. I am going to take you with me ! '

He drew a heavy pair of gloves from his pocket, which he

always carried for emergencies. They were made of double native-tanned leather, and as the gauntlets reached to his elbows they protected his hands from almost anything. He drew them on quickly and, bending down, took the cub by the scruff of its neck and pulled it to its feet.

The cub struggled feebly, snarling at its captor, but it had no strength left to resist : even so Mr Mainwaring had great difficulty in carrying it back to his bungalow.

The sight of their master squelching up the soaked path to the bungalow with a young tiger cub in his arms created quite a sensation among the servants. Not one of them welcomed the newcomer, but not for the world would they admit the fact. They shook their heads sadly and gabbled away at the top of their voices.

Mr Mainwaring strode up the verandah steps. ' Bring me a deep box, and put some straw at the bottom,' he ordered.

When the box was brought he laid the cub tenderly in it ; then he had the box carried to his own bathroom.

The cub was grateful for the warmth of the straw, for the rain had chilled him to the bone, and he did not attempt to snarl when Mr Mainwaring bent over him ten minutes later.

' I shall call you " Argh ", for it was the fire that brought you to me,' he said. ' When you feel better we are going to make friends. Now I am going to leave you alone for a bit.'

He went out softly, closing and barring the door behind him.

CHAPTER XIII

The Taming of Argh

ARGH REGAINED consciousness with a shudder. He lay still for a few moments with his eyes open. Something had happened, something was lacking ; he still stared straight in front of him in bewilderment as he gradually recovered his memory.

In a flash he had jumped to his feet, and was standing in the box, with his ears back and his mouth set in an ugly snarl. It did not take him long to realize that the dreadful roar of the burning forest, the choking smoke and the flying sparks were a thing of the past. Although by no means satisfied, he sank down into the box again and started to inspect this strange new lair.

His gaze travelled round the whitewashed walls of the bathroom, and then down to the concrete floor. He sniffed the sides of his box carefully and decided that he didn't like the smell at all, and that he had better leave it at once.

This time he dragged himself slowly up to a sitting position, testing his limbs as he did so in a luxurious stretch. At any rate he was sound in limb, and but for the smell of smoke on his coat and one or two sore spots where sparks had landed on him, he was unhurt.

Now that he knew that his legs were in working order he stepped out of the box and started to explore. He was terribly thirsty, but as he looked round him he felt his heart sink. This did not look the kind of place where water was to be found.

'No grass anywhere,' he thought. 'There is always grass round water—but the ground here is so cold—so beautifully cold.'

He crossed the room to where a little wall some two feet high enclosed a square space. The cub skirted the wall in the hope of finding an opening, but when he did find it he drew back quickly. In the middle of the enclosure there was a beast—a grey beast—standing very still.

'It is the *colour* of elephant,' he thought to himself, as he sat watching the grey, zinc bath-tub with an anxious eye ; 'but it is certainly *not* elephant. I will wait until it moves.'

Half an hour, an hour, an hour and a half passed, and still the thing did not move. The cub's patience was exhausted.

He rose to his feet and, with his tummy touching the ground in the correct stalking attitude that his mother had taught him, crept towards the bath.

'Which is the head ? ' The doubt brought him up short, and he eyed his quarry in bewilderment. Never had he seen a beast exactly the same at both ends.

'I will spring on to the middle of its back,' he decided at last, 'and bite deeply into its body.'

He tested his muscles for the charge and then sprang with

a roar—a roar that was drowned by a terrific splash as he landed
into eighteen inches of cold water. The bath skidded along the
stone floor with the shock of his charge, and crashed against the
far wall of the room.

Spluttering with rage, Argh jumped out of the water as
quickly as he had jumped in, and, without waiting to go through
the preliminary movements prescribed for young tigers about to
fight elephants, launched a furious attack upon the unresisting
sides of the bath.

It was an unsatisfactory battle. His teeth slipped off the
zinc with a grating scrape that made his very hide creep. And
when after two minutes of furious assault his enemy showed no
signs of resistance, and still seemed unharmed, he sat back on
his haunches and regarded it curiously, with his head on one
side.

' Have I killed it ? ' he wondered.

He licked his paw, and rubbed it over his nose, which had
been badly bumped when the bath had charged into the wall.
Again his amazement increased, for his paw was wet—and wet
with water ! Then he discovered that the ground on which he
was sitting was also wet and that there was a circular pool in the
middle of the stone floor—could that soft swishing sound, that
had come from the inside of the enemy, be water ?

His anger died within him. He approached the bath again
with an air almost of apology, and he was glad to see that it had
not taken offence. Very cautiously he placed his paws on the

edge of the tub and leant over, but he had splashed the water about so much that he was unable to reach it from the side. For the second time, he got into the bath, but this time without a splash, and drank deeply.

He had never tasted such excellent water before ! Compared with the shallow, sunheated pools, muddied by constant use, to which the tigers used to resort, it was nectar.

His thirst quenched, he rolled over and over in the bath in order to get rid of that smoky smell of the fire. Then, much refreshed, he got out and sat down on the floor to lick himself dry. The action brought back the memory of his mother, for she had only just taught the cubs to perform their own toilet, and a passionate longing for her arose in his heart.

' Mother ! Mother ! ' he called again and again, pausing between each cry to listen for her reply ; but strain his ears as he might he could hear no answering roar.

Suddenly he heard another sound—the sound of heavy, even footsteps ! The cub pricked his ears to listen. Nearer and nearer came the steps, right up to the bathroom door : then the door opened and Mr Mainwaring slipped into the room.

The cub leapt to his feet and faced this new enemy with every hair bristling, and for a few seconds man and beast regarded each other without moving. Then the cub moved a little to one side and glanced behind the man.

' No tail—it must be man,' he thought, ' but it does not smell like the man mother showed us. This creature smells quite differently—he smells of meat—he is a meat-eater, even as we are ! '

This startling discovery shook him considerably. He dropped back a few paces and examined the front view of his visitor once more.

' He is not black and shiny, with white patches—but it *can't* be anything else,' he decided, as he ran his eye over Mr Mainwaring, from his feet to his head. As the only men that he had seen so far were Indians wearing a white loincloth and turban as their sole clothing, it was not surprising that the cub was puzzled ; and the fact that the natives of the plains are vegetarians, living almost entirely on rice, had given them a ' distinct ' smell of their own. Nevertheless Argh had no doubt in his mind that this was ' man '.

He drew back his lips. ' Ma—an ! Ma—an ! ' he snarled.

' So *that* is what you think of me,' said Mr Mainwaring gently. ' I'm sorry about that, Argh, but perhaps I shall be able to make you change your opinion.'

' So *that* is the noise " man " makes ! ' thought the cub. ' It's not what I expected. Perhaps he only makes the great fire noise when he is angry ! '

' I am going to keep you here,' Mr Mainwaring went on. ' I would let you go if I thought that there was any chance for you, but you are too young to hold your own in the jungle—you will just have to make the best of a bad job, old chap.'

Argh listened with his head slightly on one side. Man's

roar was as soft as the wind in the trees—what was there about him that had filled his redoubtable mother with such fear ? ' They kill for pleasure,' she had said—' and give pain for pleasure too '—' Ma—an ! ' he snarled again, but there was no conviction in his voice.

What was there about this curious coloured, quite enormous man that made him doubt his mother's most solemn warnings? He came close to Mr Mainwaring's boots and sniffed at them suspiciously. Then, greatly daring, he reared on his hind legs and, with his forepaws braced against his captor's legs, gazed into his face.

' Well, you're a plucky little chap, anyhow,' said Mr Mainwaring, as he looked down at the little fluffy head. ' I wonder if you have guessed how fond I am of all you wild people—yes, even the ones that you think exist only to fill your little tummy ! '

Argh dropped to the floor again : his brain felt quite addled. Here he was, shut up in a small lair with a MAN, and nothing unpleasant had happened !

' Never will I forget what my mother told me,' thought the cub to himself. ' *Never* will I touch man—but if *this* man is dangerous then I am no wiser than a blind kitten ! '

He rolled over on the floor at Mr Mainwaring's feet, and with one paw patted playfully at the toe of his boot.

★   ★   ★

Mr Mainwaring soon confirmed Argh's first opinion of him —he certainly was, as far as the cub was concerned, harmless. But soon Argh made another, and even more pleasant, discovery.

For several days the cub lived in the bathroom, visited only by Mr Mainwaring. At first he was only interested because Mr Mainwaring brought food to him, but one day he found himself looking forward to seeing the man for his own sake.

' I must be bewitched,' thought Argh. ' Here am I, the cub of the biggest tiger in the jungle, watching that door as a jungle-cat watches a rat hole. What would my mother say ? I ought to burn with hate—but I don't—ah ! there he is ! '

Argh jumped down off the little wall and moved over to the door, his ears pricked and his tail swinging gently to and fro.

' Well, old man, pleased to see me ? ' asked Mr Mainwaring, as he entered the room. He bent down and ran his hand over Argh's head. ' I think it's time you left this room and came out with me. Don't you think that is a good idea ? '

Argh did not understand the invitation, but he saw that the door, which Mr Mainwaring usually shut the moment that he was inside, was standing ajar, and he looked towards it enquiringly.

' Yes, out through that door—but you must let me put this collar round your neck first.'

Mr Mainwaring took a bulldog's heavy collar from his pocket as he spoke, and held it out for the cub's inspection.

Argh smelt it. It was tough, but there was a tang about it not unlike the hide of a buck. He was not hungry enough to eat it then, but he might make a cache of it to fall back on in time of need. He took it in his mouth.

' No, it's not to eat,' laughed Mr Mainwaring. He bent down and slipped the collar round Argh's throat.

' Strange are the ways of man,' thought the cub, shaking his head as the big collar was fastened in position. ' Whatever *is* it for ? '

He was not kept in the dark long, for Mr Mainwaring slipped a leash through the ring of the collar and led Argh out of the bathroom on to the verandah.

Argh blinked at the strong sunlight. He had got more or less accustomed to being awake during the day, for Mr Mainwaring kept popping in and out of the bathroom, and that, combined with the noises going on in the bungalow, had made it impossible for him to sleep—but this bright light was another matter. For a moment he thought that he was frightened and drew back, straining against the collar—when suddenly he saw something that drove all thought of fear from his mind. It was a MAN—a *real man* such as his mother had shown him in the jungle.

' Grrh—MAN ! ' he snarled furiously, his hair bristling all down his back.

The *kitmagar*, who was bringing his master a drink, beat a hasty retreat, but a shout from Mr Mainwaring brought him back reluctantly. ' Put that drink down and come close,' ordered his master.

With leaden feet the man obeyed, eyeing the cub with intense disfavour. ' Sahib, see the striped one does not like me,' he protested in a quivering voice.

' Then the sooner he learns to do so, the better,' answered Mr Mainwaring. ' Now Argh,' he said, tightening his grip on the collar, ' you have just got to learn, right away, to treat my servants as you treat me.'

' He is angry ! ' thought the cub. The grip on his collar and the tone of Mr Mainwaring's voice were as firm as the commands which his mother had given him—commands which had to be obeyed. Argh's coat ceased to bristle—he had learnt the lesson of obedience from a far sterner teacher than Mr Mainwaring.

' They must be blood brothers after all, in spite of their colour and their smell,' he thought to himself. ' *And* mother told me *never* to touch the creatures—'

He bent his head and smelt the *kitmagar's* bare feet.

CHAPTER XIV

Fame

EIGHTEEN MONTHS passed—months in which Argh grew into a magnificent specimen of six feet of striped quicksilver. He was lightly built, for his muscles had not yet fully developed, but his brain was as keen as that of a full-grown tiger.

Tigers' brains, like those of human beings, vary in each individual, and Argh learnt far more in a short time under Mr Mainwaring's tuition than he would have learnt in his whole life as an ordinary wild tiger of the jungle.

From the very first day when his master led him out of the bathroom Argh laid himself out to charm the strange creatures among whom fate had brought him ; before many months passed he was welcomed joyfully wherever he went. Coolies would come in from outlying villages merely to catch a sight of Mr Mainwaring and his new companion—for the chance of seeing a white man on a horse, with a nearly full-grown tiger trotting at his heels like a dog, was not a sight to be missed. But in course of time Argh's fame spread until the news of this strange companionship appeared in a paragraph in the *Statesman* and reached to every part of India. Mr Mainwaring was overrun by visitors, who, undeterred by the terrible state of the roads, came to see for themselves whether the stories they had read were true or not.

Argh thoroughly enjoyed this new life. He had now grown accustomed to being awake for the greater part of the day, and he found man to be both a harmless and amusing beast. There was a rickety cane sofa on Mr Mainwaring's verandah which the cub made his own. Here he would lie and watch his admiring visitors, his eyes gleaming with lazy amusement, while the sofa creaked and groaned with every movement of his body. But if his master should leave the verandah for a moment he would spring from the couch like a rocket and follow him. Mr Mainwaring was the only man whom he really trusted, and he would not willingly let him out of his sight for a moment.

But at night things were different ! As soon as everybody in the bungalow was asleep Argh would wander off into the jungle. The fact that the door of the bathroom in which he slept was left open at night was a secret known only to the cub and his master. It was Mr Mainwaring's wish that one day Argh should return to his own people, and he realized how vital it was that the cub should retain his jungle sense, and should be able both to protect himself and also find his own living. Now that he was grown up, therefore, Argh was only fed in the morning, and as a result he was often hungry when he slipped out of the bathroom into the darkness, to spend the night tracking his dinner and making his kill.

One morning Argh, comfortably full of a spotted deer that he had killed as neatly as anything his mother had hunted, sprawled on the verandah, watching his master at breakfast. By Mr Mainwaring's side lay an immense pile of letters, for his post had grown considerably since the coming of Argh.

The cub eyed the pile with distrust. ' Will they make the MAN angry this morning ? ' he wondered. ' There is some bad magic in those things—he growls at them, tears them in pieces and casts them away—but it does not stop them. More and more come, like bees in swarming time.'

Mr Mainwaring finished his breakfast and then turned to his letters with a sigh. He slit open the envelopes and scanned the contents. ' Three more newsrags wanting photographs and descriptions of you, Argh,' he said. ' You are growing as popular as a cinema actress—six Nosey Parkers wanting to spend a week-end with us—what do they think this bungalow is—a hotel ? Two bills for photographic material—hum, less than I had expected.' He picked up the last envelope on the pile and looked at it closely before opening it : then he slit the flap and glanced at the letter inside. Argh saw his face cloud with annoyance, and climbed on to the sofa to be nearer to him.

Mr Mainwaring read the letter through and then swore forcibly : ' Argh, my lad, you have grown a jolly sight too famous. Here's a fellow offering me three thousand rupees for you ! He wants to train you to perform in a circus—of all abominable things ! '

Argh threw back his head and yawned, making his seat creak ominously.

' It's all very well for you to yawn ! You know that I wouldn't sell you for all the wealth in the world—and to a circus of all things—but once it gets known that somebody is ready to pay two hundred and twenty-five pounds for you there are plenty of people who will only be too glad to have a shot at kidnapping you ! '

Mr Mainwaring looked at the tiger with a smile. ' The sooner you set up house for yourself in the deepest part of the jungle you can find the better I shall be pleased—that is, if you don't break your neck by falling off that couch first ! '

Argh yawned again. ' Why doesn't the man kill the thing and throw it away ? ' he thought.

He got off the couch as Mr Mainwaring rose from the table, the letter still in his hand. Argh eyed it warily, and then, as his master bent to stroke his head, snatched the paper from his hand and, placing one paw upon it, tore it into two pieces.

Mr Mainwaring laughed. ' No, it can't be dealt with in that way ! ' he said, as he recovered the letter. ' And in any case the man says that he is on the way here now to talk about it —curse him ! '

He paused and ran his hand across his forehead. ' I wish I could shake off this malaria—it looks as if I am in for another bad " go "—well, I hope the mosquitoes bite our enterprising friend so that *he* gets a dose—that would teach him to come blundering into these forests on a wild-goose chase ! '

He cuffed Argh playfully on the head, and the tiger, swishing his tail slowly, registered his complete agreement.

Mr Mainwaring watched him with approval, admiring the beautiful sleekness of his coat, and the rippling muscles of his shoulders. ' One blow of that paw of yours, old boy,' he murmured, ' would do more than a hundred mosquitoes to put our friend out of action ! '

CHAPTER XV

Fever

WHEN MR FRANK BENSON arrived at Barikhot station after a long, hot and tiring journey from Calcutta, he imagined that the worst of his journey was over, and the information that a hundred odd miles of almost impassable jungle roads separated him from his destination came as a nasty shock. So far his one experience of jungle life was a fleeting glimpse of a wild boar scurrying into a patch of cane—but he was not to be deterred.

He was a fat little man and had already endured much, but under those layers of fat beat a heart that would not admit defeat.

He stood sweltering on the sunny platform, surrounded by his luggage, which included an enormous cage, and argued with the station-master. ' If I cannot go by car, then how do I get there ? ' he enquired.

The Parsee station-master, longing to continue his interrupted doze in his cool back-room, eyed Mr Frank Benson with disfavour.

' Sahib, you can take a bullock-cart, ride upon the back of a horse, or—' he ran his eye maliciously over Mr Benson's generous figure—' you could walk.'

' I cannot ride a hundred miles in one day,' persisted Mr Benson, ignoring the other two suggested modes of transport. ' Is there a hotel on the road ? '

A smile creased the corners of the station-master's mouth.

' No, Sahib. There are no hotels in these parts,' he answered.

' Then *how do* I get there ? ' Mr Benson's voice was becoming high-pitched with irritation. ' Isn't there someone in this god-forsaken spot that can tell me ? '

The station-master sighed. ' But, Sahib, I have already told you. In a bullock-cart, on the back—'

' Bonehead ! ' Mr Benson almost screamed. ' Where can I sleep on the way ? Where do I get a horse ? How—'

' Can I do anything to help ? ' The voice cut across Mr Benson's tirade and pulled him up short. He swung round and approached the newcomer, furiously mopping his dripping forehead with a large silk handkerchief.

' You can indeed, sir ! I wish to go to Mr Mainwaring's bungalow—I must see him on most important business—and this dolt,' he indicated the station-master with a wave of his hand, ' refuses to tell me how to get there ! '

The station-master caught the eye of the newcomer and shrugged his shoulders.

' Doctor sahib, I have tried to tell him, but—'

' All right, station-master, I will deal with this.'

As the station-master faded quickly into the background the doctor turned to the perspiring Mr Benson and took him by the arm. ' Won't you come along to the club with me and have a drink ? ' he suggested soothingly. ' My name's Cartwright— Major Cartwright—I'm the Civil Surgeon here.'

Mr Benson allowed himself to be led to the doctor's car, and later, with a whisky and soda before him, explained his troubles.

' Do you know Mr Mainwaring's tiger ? ' he began.

' Know Argh ! I should jolly well think I do ! He's a great pal of mine,' rejoined the doctor.

' Well, that's what I have come about. I have offered Mr Mainwaring a handsome sum—a *very* handsome sum—for this tiger of his, and I have come to collect him and take him back with me.'

' You don't mean to say that Mr Mainwaring has consented to *sell Argh* ! ' The doctor's voice sounded thunderstruck.

' No, no ! He has not had time. I wrote to him last Saturday, and followed my letter immediately. But there is no doubt that he will accept—no doubt at all ! Any man in his senses would jump at such an offer as I have made.'

' I don't know so much about that,' said the doctor. ' Mainwaring is devoted to the tiger, and in no need whatsoever of money—lucky devil ! '

The doctor's voice sounded very definite, but it in no wise shook the confidence of his listener.

Mr Benson laid his finger on the side of his nose, and winked knowingly. ' You wait and see,' he said. ' He'll sell all right !

The doctor suppressed his irritation. 'Very well,' he said. 'If you are so sure about it you had better come along with me. I've got to go into the jungle myself, for I have several patients to see.'

★   ★   ★

Argh was worried. The magic in that letter seemed to be even more potent than any that had come before, for after reading it his master had been stricken down by some invisible power. He now lay on his bed, tended by a demented doctor-babu.

'A runner must go at once to Barikhot and fetch the doctor-sahib,' he babbled to Mr Mainwaring's servants, who stood huddled in the verandah, dazed by the catastrophe that had suddenly overtaken them. 'I cannot tell what it is that ails him —he said yesterday that he had a touch of malaria, and I have treated him—but this fever—it is not malaria ! I cannot tell what it is ! '

'The doctor-sahib is on his way here, babu.' It was the dispenser who spoke. 'He is coming to see—'

'Oh yes ! Oh yes ! I remember ! ' interrupted the doctor-babu. 'The gods be praised ! All this worry had driven it from my mind.'

'Of course, of course,' echoed the servants with relief. Usually the Civil Surgeon's visits had been of a social nature, and in their agitation at their beloved master's danger, they had forgotten that he was about to make one of his tours of inspection.

' All will now be well,' they assured each other optimistically. ' The doctor-sahib will heal him.'

But there was one person who did not share in their relief. Argh knew nothing of doctors, and continued his walk up and down the verandah outside his master's room. A sensation that he had never experienced before seemed to be turning his blood to water. Had the sickness got him too ? Up and down, up and down he walked like a caged animal, every now and then uttering a low growl in the effort to ease the torment of his mind.

Twice the servants had tried to entice him away, but the savage snarl that greeted their efforts convinced them that they had better leave Argh alone if they valued their skins.

CHAPTER XVI

Stolen—But Paid For

MAJOR CARTWRIGHT arrived at the bungalow alone. A man had been sent to meet him with the news of Mr Mainwaring's illness, and, after a strongly worded hint to Mr Benson to return to Calcutta, the doctor had galloped on ahead.

His face was grave when he came out of the bedroom. ' Typhoid, I'm afraid, doctor-babu,' he said. ' He's in a bad way, but with a constitution like his he ought to pull through all right. I'd like to get him into the hospital, but that is impossible. We must just do the best we can here.'

The babu nodded. ' He shall have the best attention I can give him,' he answered, in a choked voice. ' We—we peoples round here hold him in high esteem.'

' Of course, doctor babu. I have every—what the— ! ' Major Cartwright broke off and stared at his late travelling companion, who had just appeared in front of the bungalow, with a cold and furious eye.

' I thought I told you to go back to Calcutta,' he said icily. ' Mr Mainwaring is dangerously ill.'

Mr Benson took not the slightest notice. For a few seconds

he sat on his horse and watched the unhappy tiger as he paced up and down ; then, without a word, he turned his mount and rode out of the compound.

'That's got rid of him,' thought Major Cartwright with relief, and turned back to the doctor-babu.

*   *   *

But Mr Benson had not gone far. Without much difficulty he found the house of one of the overseers of the labour gangs, and by dint of persuasion and the promise of a rich reward, he induced him to provide some place where he might sleep. He had decided to lie low until his luggage, and especially the cage, should arrive from the station on a bullock-cart.

As soon as he had completed his inspection Major Cartwright was forced to return to Barikhot, and the little doctor-babu, fortified with advice, was left to cope with the situation single-handed.

It did not take Mr Benson long to persuade the anxious little man that Mr Mainwaring had really intended to sell the tiger to him, and the suggestion that the tiger should be removed was a great relief, for Argh's ceaseless parade on the verandah was playing havoc with his nerves, strung as they were to their last pitch by the great responsibility which had been thrust upon his shoulders.

'But—but have you the money with you ? All that money ? ' he enquired, when Mr Benson announced the price that he was ready to pay.

' I will pay it in *cash* ! You shall receipt the bill now, and you can pay the money over to Mr Mainwaring when—and if— he recovers.'

The doctor-babu gave in. The fate of Mr Mainwaring was of so much greater importance that it did not seem worth while to make a fuss about such a transaction—especially as Mr Benson assured him that it would have already been completed but for Mr Mainwaring's illness.

' But how will you catch him and put him in your cage ? ' inquired the doctor-babu politely.

' Don't you worry—I will attend to that,' Mr Benson answered confidently. Argh's feelings in the matter were not consulted.

No sooner had the money been paid over than Mr Benson set about his plan for securing his newly acquired property. As a preliminary the cage was trundled up to the bungalow, and placed on the verandah with Argh's dinner inside it.

Argh paid not the slightest attention to these preparations, but continued his walk up and down, up and down, outside his master's room. From the room he could hear Mr Mainwaring babbling in a voice that he scarcely recognized. ' Has he been bitten by the monkey people ? ' he wondered, as he listened to the ceaseless flow of words. ' None but the monkey people chatter like that—chatter that pleases none but themselves.'

Mr Benson waited patiently. Just as he was beginning to fear that the tiger was going without food that day, Argh became suddenly conscious of hunger. He looked about him, lashing his tail. 'Where is my dinner? ' he snarled. 'Why is it not in its usual place? I can smell it—where is it? '

Mr Benson interpreted the snarl aright, and waited in a fever of excitement as Argh, his hair bristling, approached the cage. For a moment he feared that the tiger would not enter the cage, but, since he had fallen under Mr Mainwaring's spell, Argh had learnt not to fear an enclosed space. Although surprised that his dinner should have been placed in the middle of a clump of dried bamboo—for he smelt the new cage carefully and noticed only the ordinary ' man smell ' to which he had become accustomed—he entered the trap with confidence. As he seized his food the door behind him dropped with a clang !

All his mother's warnings came back to Argh with a rush. He spun round with a roar of fury and tore at the door with teeth and claws. But even as he leapt at it he lost his balance and rolled over on the floor. The cage had been snatched up by the men whom Mr Benson had hired in the village, and was now being trundled down the steps and across the compound. Hurled from side to side, and unable to get a foothold, Argh called with what breath was left to him.

' MAN ! MAN ! '

Even if Mr Mainwaring had been in a condition to hear it, he would scarcely have recognized that gasping call for help.

Mr Benson imagined that it was a cry of pain. ' Gently there, gently there, you clumsy brutes ! ' he yelled to the carriers. ' Can't you see that you are hurting him ? '

In his heart Mr Benson knew that he had been guilty of sharp practice, and it was not until the cage had been hauled through the jungle, and placed safely on the train bound for Calcutta, that he breathed at all easily. He was glad that there had been no sign of Major Cartwright at the station.

'At any rate I have paid for the animal—and a very good price too,' he muttered to himself in justification. 'His late owner will have every reason to be pleased, *if* he recovers—and *if* that little ass of a babu hands over the cash.'

But there Mr Benson—who was no judge of men and only appraised animals by their cash value—was at fault. For the little ass of a babu, by dint of a day-and-night struggle lasting for three weeks, pulled his patient round ; and, being a man of the most rigid honesty, handed over the thirty notes, each value for one hundred rupees, to Mr Mainwaring at the first sign of convalescence, in the hope that the sight of all that crisp money would cheer him up.

It was no doubt unfair that the storm, which should rightly have broken on the head of Mr Frank Benson, now safely on the high seas, beat against the doctor-babu and reduced him to tearful incoherence. For the first time during his service in India Mr Mainwaring, who was in no fit condition to hear that Argh had been removed by a trick, lost his temper. His rage was so great that he fell into a relapse, and the doctor-babu's work started all over again. By the time that Mr Mainwaring was well enough to do anything, all hope of recovering Argh had to be abandoned.

CHAPTER XVII

Through the Jungle

As MR BENSON watched the bullock-cart swaying along the rough track which wound through the forest to Barikhot he breathed a sigh of relief. It had not been easy to explain to the coolies exactly what he had wanted done, but he had in the end succeeded by sheer force of will-power. The gang overseer, who, fortunately for Mr Benson, spoke a little English, had been alternately bribed and cajoled into finding a bullock-cart owner and coolies willing to make the long journey, and at last the wretched Argh found himself jolting along on the first lap of his terrible journey to England.

A less experienced man would have gone ahead and waited for the arrival of the bullock-cart at Barikhot, but Mr Benson had not spent his life among captive wild animals and their attendants for nothing. He knew the constant delays and troubles that arise in moving a menagerie from one country town to another, even over good English roads, and he was determined not to let Argh out of his sight, no matter what discomfort such a decision might entail.

It was with a heart full of self-satisfaction that he mounted his horse to follow in the wake of the cart. Slightly behind him, perched on a miserable specimen of bazaar pony, rode his servant, a hired groom from a racing stable in Calcutta—a man as unused to discomfort as his master.

Ali hated the whole business from the bottom of his heart, and would have deserted Mr Benson long ago if that astute gentleman had been foolish enough to pay him his wages in advance. As it was he was compelled by lack of cash to stay with his temporary employer, but he promised himself that this was the last time he would ever forsake the comfort and excitement of Calcutta in order to go on a madcap excursion into the jungle, no matter what the bribe might be.

As he rode behind Mr Benson, Ali turned over his grievances in his mind. For four days he had been housed in a squalid village, occupying quarters that he would have considered almost unfit for a sweeper, mocked and despised by the jungle people, and treated by the very coolies as a man of no account—he who had once ridden two winners in one afternoon on the famous Calcutta racecourse ! Every bone in his body ached with a terrible stiffness after the long ride from Barikhot on the back of the unspeakable animal that had been provided for him, and now he was facing the same journey again on the same beast— and all for the sake of a tiger ! As if there weren't hundreds of tigers that could have been procured without all this trouble ! As he shifted about on the rough felt saddle, trying to find some part of his anatomy that did not feel as though it had been pounded to a pulp, he consigned Argh to the nethermost regions.

But Argh, the unwitting cause of Ali's troubles, was suffering even more than Ali's vindictive soul could wish. As his cage swayed backwards and forwards on the bumpy bullock-cart, he felt absolutely miserable in mind as well as body. It was not fear or pain that moved him—though he had been through enough to frighten the stoutest of tigers, and every jolt of the cart caused some splinter of bamboo to jab him in the ribs or belly ; he was miserable in mind and spirit. Man—the great friend in whom he had trusted—had failed. Man—the hairless monkey that he had spared only at the great friend's bidding—had shut him up in a cage, poked food at him with a pole, and howled and chattered at him !

Through the bars of the cage he could see the leafy shades of the jungle—his jungle ; he could sniff the smell of earth and forest that he loved. In the cool of the evening he could hear the jungle sounds that he alone of all that melancholy procession could interpret : once he winded the game he had so often stalked—a herd of deer gathered about a water-hole, that stopped to listen to the cart creaking along the jungle track.

When the bullock-cart halted for the night, and the coolies started to make camp, Argh made a frantic effort to free himself, but his teeth and claws could no more than scrape the bars— there was not room enough for him to gather for a spring and throw his weight against them.

Night was far worse to bear than day. The sounds and scents grew stronger and stronger, and he raged helplessly ; now snarling with the anger of a hungry beast mauling its kill, now sending a challenge, a deep-throated roar, echoing through the lofty trees. It might have been some comfort to him had he known that the noise he made was the last drop in Mr Benson's cup of discomfort. Whenever Mr Benson heard that roar he shivered with rage and cursed—suppose the beast could summon other tigers to its aid ! He bade the coolies build a great fire, in spite of the clammy, airless heat of the jungle that seemed to cling to his throat and stifle him.

Worn out by the long day's ride the circus proprietor had crawled gratefully on to the camp-bed erected by an unwilling Ali—but owing to the haste with which he had left Calcutta he had forgotten to bring a mosquito net, and Ali was not the kind of servant to bother about such a trifle !

Mr Benson's head had scarcely nestled on to the pillow before a whining cloud descended upon him. Ping ! ping ! Bzz !

Sleep was well-nigh impossible ; even when he imitated the coolies and drew his sheet over his head he was in no better plight, for he could hardly breathe, and the sweat poured off his face and body.

By three o'clock in the morning Mr Benson could stand no more, and Ali, who had spent a night of sheer terror, expecting at any moment to be carried off by a leopard, was only too thankful when the order came to move. By their united efforts they succeeded in waking up the men and getting the cavalcade on the road once more, in spite of the bitter complaints of the overseers and coolies.

'The bullocks are tired—they must rest!' protested the driver of the cart, who, rolled up like a heap of dirty rag, had been sleeping in peaceful oblivion—even Argh's roars had not penetrated his dreams.

The forced march that followed would have done credit to Hannibal. With only short snatches of rest for food and sleep Mr Benson drove them on throughout a whole day and night, shouting and gesticulating, and finally using a stout bamboo *lathi* to goad the unwilling men to take the track again once they had fallen out. As a result, by six o'clock the following morning a jaded, dusty, red-eyed, but triumphant little man, a good deal slimmer than the prosperous gentleman who had descended from the Calcutta train a week ago and fussed about as if he were a person of some importance, stood on the platform at Barikhot enquiring of a sleepy station-master when the next train left for Calcutta—and all for the sake of Argh!

Luck was with Mr Benson. Within an hour he was stretched upon his back in the bunk of a saloon car, sleeping like a log, while the train whirled him back to civilization.

But that train journey brought anything but relief to poor Argh. For the second time in his life he experienced panic—blind, unreasoning panic—and this time he seemed to be in a far greater plight, for when the jungle burned he had been able to make some effort to save himself, but now he was helpless, penned in a tiny cage where he had only just room to move.

The world rocked and screamed about him, and, except when the train stopped at the stations, he could hardly hear the sound of his own roars of rage and terror.

CHAPTER XVIII

Captivity

MR BENSON had been so confident that he would bring Argh back to Calcutta that he had made full arrangements for the housing of the tiger before he left for Barikhot. There was an empty cage in the tiger-house at the Zoo, and Argh was carefully ushered into it early one morning after the arrival of the train at daybreak.

The cage had been hoisted out of the closed goods truck on to a motor-lorry, and within an hour Argh was stretched out in the sawdust on the floor of his new quarters—too tired and too weary in mind even to finish off the chunk of meat which had formed the bait to lure him from his travelling cage.

For a fortnight he lay there listlessly—well fed and looked after, but heeding little of what was going on around him.

He was conscious of the smell of other tigers on each side of him, and during the first night he had scratched against the partition separating him from them, roaring and challenging them to fight. They had been bred in captivity, and it would have gone ill with them if the new arrival could have got through to them—but they did not know this and added to the din and clamour by roaring their defiance in return.

It had been Mr Benson's intention to make a public show of Argh in Calcutta, so that the tiger might enhance his reputation before being shipped to England, but he now decided that it would be wiser to attract as little notice as possible. But the secret of a new arrival at the Zoo—and such a magnificent specimen—could not be kept long.

The officials were loud in their praises when they first saw Argh, for he was in fine fettle in spite of his sufferings during the last week. He had now reached full growth and was almost half as large again as his neighbours ; his coat was in splendid condition, clean and glossy : his muscles, when relaxed, rippled beneath his skin : his teeth were milk-white, and there was about him that alertness and intelligence that can only come to an animal that has been trained to fend for his own living.

At first, while he lay listlessly on the floor of his cage, these points were barely noticeable, but as he grew used to his surroundings his curiosity got the better of him, and he would prowl round his cage with every sense alert.

The coloured saris of the Indian women, and the gay clothes of the men, who flocked to see him, excited him at times almost to a frenzy, so that Mr Benson began to wonder if his new capture would ever get used to the glare, colours and noise of the circus tent. On the whole it was a much relieved little man who superintended the hoisting of Argh's cage on board one of the large cargo steamers that ply from Calcutta to London.

Mr Benson had made arrangements with the shipping company for a passage for himself, and he did not leave the cage, which had been stowed forward on the lower deck, until the ship had cleared the port and was half-way down the Hoogli. He kept one eye nervously on the teeming quayside, fearing that even at the last minute Mr Mainwaring or his representative might turn up and claim the tiger.

Every stranger that came aboard sent a shiver down his back—but all was well, and when the last siren blew, the last hawser had been cast off, and the ship started to move slowly through the water, he heaved a sigh of relief that the much coveted Argh was at last his undisputed property.

Argh had been frightened when his cage was hoisted up by the derrick—but he plucked up his courage at the sound of the siren and roared back his defiance. The movement of the ship was barely perceptible, while the sound of the engines turning over slowly as she ploughed through the river seemed just a gentle throb—a throb that the unhappy tiger found strangely comforting. It reminded him of the sound of distant tom-toms, being beaten after dusk in the villages surrounding the jungle.

Argh eyed Mr Benson without interest as the little man settled himself down on a stool beside the cage. ' What does this man want with me? ' he wondered. ' He has followed me ever since I entered that box—is there no escaping him? '

Mr Benson, in spite of the relief in his mind, eyed his captive with considerable concern. Argh was to be the star turn in his circus and yet—would it ever be possible to tame his spirit? Had he not made a mistake in taking a young tiger that had experienced jungle life at its fullest, even though he had lived with Mr Mainwaring for nearly two years ?

At the moment the tiger's appearance suggested that he would achieve greater success with the public as a hearthrug than a living animal, so dejected did he look.

Mr Benson leaned forward towards the cage, and spoke in a low, even voice. ' Come, come ! ' he said gently. ' You must buck up ! Think of your future ! I have ruined my holiday for your sake so that you may be the craze of England—the most famous turn in a circus known throughout the world for its brilliance ! Is that not better than being buried in the back of beyond ? '

Argh opened his eyes, which had been closed in utter boredom and despair. This man was not so bad after all ; he talked in a pleasant, soothing voice—he did not shout, or wear red, yellow and green clothes.

Mr Benson was quick to notice the tiny spark of interest and he went on talking—as the ship slipped down the river he revealed his plans and hopes for the future ; he spoke of the glories of the sawdust ring, the lights and laughter of the children and the inspiring thunder of applause. ' From the moment I read of you in the papers I knew that you were the tiger I had been looking for for years. I, Benson of Benson's Mammoth Circus, cancelled all my plans for my holiday in India to rescue you from your backwoods to give you the publicity you deserve ! '

Argh, his eyes now glued on the fat figure on the camp stool, settled himself more comfortably on the floor of his cage. He knew men too well to be afraid of him as a man, and even the purring drone of Mr Benson's voice pleased him. He was sorry when at last the man rose, folded his stool and disappeared from view. His heart ached for Mr Mainwaring who was to him the ' MAN ', but this was undoubtedly one of the same breed—he had the same colour and the same gentle way of talking.

Mr Benson, seated in the small saloon with a much needed drink before him and a fan buzzing above his head, congratulated himself on his budding friendship with the tiger. ' I'll have him so tame by the time we get home that they can take my photograph with my arm round his neck—and that will fetch the press ! ' he boasted to the chief officer. ' I'll spare no effort to get on the best terms with him before we land ! '

The mate smiled. ' There was a young lady of Riga,' he murmured. ' You had better not put on a skirt the next time you go and see him ! '

Mr Benson never let bodily discomfort interfere with anything connected with his business, and the winning of Argh's friendship demanded a lot of hard work. It was terribly hot in the Bay of Bengal, and he was not a good sailor, but not even sickness turned him from the task that he had set himself.

Argh also suffered from the heat, in spite of all that Mr Benson did to relieve it, but though his body was acutely uncomfortable, his mind gradually became resigned to his lot. He missed his happy home terribly, with its freedom from all restraint ; he longed for the constant companionship of his old master and, most of all, for those midnight huntings in the jungle.

They had put up an awning over his cage to protect him from the glare of the sun, and Argh was as comfortable as they could make him, eating little but drinking long draughts of water given to him by Mr Benson himself, who tipped lavishly in order to get what he wanted. To do him full justice the little fat man would have done as much for any beast in distress, whether intended for his circus or not.

Apart from his profession, or perhaps because of it, Mr Benson had a real affection for animals, and it was merely lack of imagination, and an overweening opinion of the honour which his circus bestowed upon its inmates, that made him unable to realize that the glory of a public career might not compensate an animal for the loss of its freedom.

He knew that they did not like it at first, but he would never have admitted that, once broken in, they could find the life anything but perfect under his charge.

After a passage so calm that it was difficult to believe that one was at sea, the ship docked in London, and Argh was hurried away to Mr Benson's private house, where he was put into a converted stable, with a small yard for exercise. The change from the cramped existence in the travelling cage to the large and airy quarters put new life into him. He forgot those awful hours of blinding heat when the ship steamed up the Red Sea with a following wind ; he forgot the noise and clatter and dust of coaling at Port Said. He grew fit quickly, assumed his old air of alertness and intelligence, and even began to look forward to Mr Benson's daily visits.

For the circus proprietor had decided to undertake Argh's training himself. It was years since he, as a young man, had started Benson's Circus with a mere handful of animals all trained by himself, but his hand had not lost its cunning. Argh could not have had a more capable, and a more kindly, trainer.

CHAPTER XIX

Trial by Jury

IT WAS in August that Argh met Bill, the official lion-tamer in Benson's Circus. There were six feet all told of Bill, topped by a black, glistening head. His large, round, red face glowed with health, and was adorned by a pair of magnificent curling mustachioes which matched his fierce shaggy black eyebrows. He looked fiercer than any lion or tiger, but actually he was one of the gentlest of men.

When Bill stood in the middle of the ring in his spotless white tight trousers and his red coat barred with gold, cracking his enormous whip, people were not surprised that lions and tigers obeyed his lightest behest as if they were a troupe of performing cats.

Bill was not dressed in all his glory when Argh first made his acquaintance—that was to come later—but even in ordinary clothes he filled the eye. Bill, who was inordinately proud of his personal appearance, imagined that he had the same effect on the animals under his control that he had on human beings, and the truth would have been a cruel blow to him. For every animal in the Benson menagerie regarded him as a joke, the best, most toothsome joke that a kind Providence had sent to brighten their dull lives.

'Now this tiger is the cleverest we have ever handled,' said Mr Benson, as he led Bill across the stable-yard one morning. 'He has brains, real brains. I told you his history—perhaps that may account for it. He'll do anything you ask him once he has grasped what is wanted, and he is as tame as a cat.'

Argh now spent most of the day in the exercising yard—a large open space surrounded by tall, iron railings, which had been curved inwards at the top so as to prevent him jumping out. A small tree grew in one corner, and gave the tiger shade when he needed it. The yard was a source of great pride to Mr Benson, and had frequently been photographed for the press ; and its owner, with justice, was held up as a model among humane circus owners.

Through the bars Argh watched the two men approaching. Mr Benson he knew—but who was this other creature ?

Argh did not take his eyes off Bill from the moment the two men had come into view, and when, just short of the cage, Bill gave a thoughtful flick to the whip he always carried, Argh reared up against the bars and roared his appreciation.

'See, he is welcoming me,' said Mr Benson, although Argh had actually forgotten his existence. For the first time since Mr Benson had taken him away from his home Argh forgot his boredom. He was amused and showed it. Round and round the cage he frolicked, chasing an imaginary prey like a kitten ; then, as though slightly ashamed of such childish behaviour, he sat down and watched Bill.

The man's every movement was a delight.   When asked to perform the tricks which Mr Benson had taught him, Argh complied willingly, uttering low growls of delight whenever Bill did something that was especially amusing.

Mr Benson and Bill put in day after day of hard work with Argh, until they, with reason, considered that they had produced a performing tiger that was absolutely unique.   It only remained to put him through his tricks before a large crowd.   As a rule this was but a small item in the training of a beast, and neither Bill nor Mr Benson thought for a moment that Argh would give any trouble—but they were much mistaken.

His first audience was a small one, consisting of the circus hands, their numerous families, and a sprinkling of Mr Benson's most intimate friends, who had been brought over to the circus proprietor's house for the experiment.

The first handful of people did not trouble Argh at all, but as more and more joined them he began to get uneasy.   A huge circus-cage on wheels, in which it was intended that he should perform, had been brought into the enclosure, and now Argh had been driven into it.   It was a gimcrack affair to look at, though strong enough.   Argh had the reputation of being the friendliest of tigers and might, in his owner's opinion, well have been taught to perform in the open, but the prejudices of local magistrates against the presence of loose tigers in their midst made this impossible.

Three sides of the cage consisted of gilded iron bars, while the fourth was taken up by a wooden door through which Bill was to enter when the great moment arrived.

Argh wandered up and down the bars on one of the open sides, eyeing the gathering crowd with apprehensive eyes. Somewhere at the back of his mind a memory began to stir. He was conscious of a feeling of intense discomfort, as if some living thing was moving inside his head. What was wrong?

Slowly the crowd gathered round, talking loudly and excitedly, and Argh's discomfort increased—then a woman raised a sunshade she was carrying, and pointed it towards the cage. Something snapped in Argh's brain—*he knew now what was worrying him.*

It was not a clear-cut memory, but it was enough. ' A closed lair—man, hairless, brown chattering man—shouting and prodding with sticks ! ' The thoughts went through Argh's head in a jumble. ' This has happened before—and I must fight ! I am afraid, but I must not show my fear—I must fight—it is my only chance ! '

With a roar that sent the little crowd scattering like a crowd of frightened hens, he charged furiously against the bars of the cage, until it rocked backwards and forwards on its wheels. Again and again he charged, tearing at the unyielding bars with frantic claws until his pads were torn and bleeding. But the

cage was made of stronger stuff than the prison in which his
mother had been confined in the Maharajah's yard, and not a
bar moved in its socket.

Mr Benson and Bill did their best to soothe him, but they
might as well have tried to calm an October gale. Until night-
fall the unhappy beast raged helplessly ; then worn out and
bleeding, he dropped down on the floor of the cage.

' I don't think we had better open the door to-night,' said
Mr Benson nervously. ' He will be calmer to-morrow.'

' You can bet your sweet life that you will never open that
door to let me *in*,' muttered Bill. ' That beast's plumb crazy,
governor ! You've been sold a pup ! '

He spat solemnly to emphasize his determination.

★   ★   ★

Argh was a failure—of that there could be no doubt. Mr
Benson spent a sleepless night working out the amount of money
he had spent on him—and in the morning he felt quite ill. So
ill was he that he departed to the seaside for a fortnight's holiday,
and handed Argh over to the head-keeper of his menagerie.
Argh was put into a small cage, and joined the row of animals of
doubtful value—animals who only appeared as exhibits, not as
performers in the circus—and could be gazed upon by those
enterprising people who were prepared to pay an extra three-
pence when the circus was over to go behind the scenes into the
menagerie tent.

In his new quarters fear and rage soon gave place to misery.

He had no idea why he had suddenly been put into a horrible little cage, why no one ever came to see him except to clean out his cage or to feed him.

All day and all night he lay on the hard floor, or paced up and down by the bars. The cage was so small that two steps took him to the end of it, and he had to turn. Soon he had rubbed a bald patch on his nose where he brushed it against the sides of the cage every time he turned.

Time had no meaning for him. One day was exactly like another, and Argh, as he lay on the floor of the cage, his eyes glazed with misery, lived once more in the past. The close-set bars on which his eyes were fixed sometimes seemed to disappear, and once more he would be back in his rightful home, free and happy. Only such memories as these made his present lot bearable, and sometimes even that escape was denied him—he could not even summon up a picture of his jungle. On such days he was filled with a blank despair, and, instead of lying on the floor of his cage, he would pace up and down in the confined space— up and down, searching hopelessly for a way of escape.

It was on just such a day that he heard an unusual noise behind his cage—the grating of a key and the clang of a bolt. The door was being opened !

Usually his food was pushed through a trap in the door, and though it was not feeding-time Argh thought that they were bringing him his dinner. He watched the door without interest, and, when it had stood open for some time without anything further happening, he strolled idly across to it and walked through the opening.

He found himself in another, even smaller cage. Annoyed and disappointed he tried to return the way he had come, only to find that that way had ceased to exist, for an iron plate dropped in front of the entrance with a clang, causing him to jump quickly away with a snarl. Almost at once the cage into which he had entered began to move.

It was quite dark, and he had not the slightest idea what was happening to him. There was no room to fight, but he had long lost any desire to fight, so he made the best of a hopeless case and made himself as comfortable as was possible. Life had become unendurable anyway and if he was to pass the rest of his days in this swaying dark hole, it was a comfort to think that those days could not last long.

But Argh's life in the swaying cage was to be a short one, and when the door at the end of the travelling box was opened again he lost no time in going through with a rush.

CHAPTER XX

Valhalla

ARGH, THOUGH he did not know it, had been sold to Whipsnade Zoo.

When the door of the little cage in which he had been travelling was opened, he saw before him a large, deep den, and he entered it willingly. The fact that a door shut behind him did not worry him at all—he had grown accustomed to that since he had left India.

He found himself in a large, airy cave, with the inevitable bars on one side only. Through the bars he watched his late abode being hauled up the steep concrete incline until it disappeared from sight. He waited until the sound caused by the removal of the travelling cave, and the noise of the motor lorry which had brought him, had died away in the distance ; then he turned his attention to his new home.

But he had no time for more than a glance around the walls when a new sound caused him to stand stock still, with his ears pricked and his eyes alight with expectancy. From somewhere quite close came the voice of another tiger !

Argh's heart thrilled at the sound. It seemed years since he had heard the voice of one of his own people, and for a moment he stood silent, scarcely able to believe his ears. Then he found his voice, and growled gently in reply.

The voice was that of Rani, a full-grown tigress who occupied the adjoining den. Argh moved over to the open side of his cage and stood close to the wall that divided him from his companion in captivity, and the tigress followed his example. They stood thus in silence for a long time—that silence between beast and beast when they communicate their thoughts to each other.

Argh sighed deeply. ' I have travelled so much—so much : I do not know if I shall stay here. I will tell you my story and you shall tell me yours.'

Rani stood silently, her tail gently swinging, as Argh told her the story of his life.

' I too have come from India,' she said, when he had finished. ' I fell into a pit from which I could not escape. Men came and opened one side of the pit and I thought that my way had been cleared for me—but I found myself in a wooden cage, even as you. I too have flown through the air—' She seemed to shiver at the remembrance of that terrifying ride through the air when she had been hoisted on board by a derrick—' but do not let us talk of such things—they are better forgotten.'

' Is there any way out of this den ? ' asked Argh, when he had explored his new quarters thoroughly. He was now lying at full length beside the bars, gazing at the sky. A feeling of peace and contentment stole over him. The smells were good—clean scents of trees and fields : sweet, pure air : bright sunshine and, above all, no noise. ' Is there any way out ? ' he repeated.

' The wall behind you rises of its own accord, and beyond are rocks and a pool. There are trees all round—but one cannot pass over to them, for the rocks are steep, steeper than any rock I have seen before.'

Argh yawned, showing a double row of milky-white teeth. ' Does man come to this place ? ' he asked, almost as an after-thought.

' Yes,' answered the tigress. ' One man comes every day when the heat of the sun is spent. The choicest portion of a kill—luscious meat in plenty. You can eat your fill for one day and rest assured that more will come on the morrow. There is no need to eat enough for several days, or to make a cache.'

' Is the man who brings the kill the only one who comes ? ' asked Argh.

' The only one who comes *close*,' said Rani. ' Others come —sometimes a few, sometimes as many as the flies around a carcase. They laugh and chatter like the monkey folk, and walk about on the road above us—'

Argh interrupted with a roar of displeasure. He stood up and lashed his tail. ' That I will not bear ! ' he snarled. ' They wave sticks—they make noises that—'

' That is foolish talk ! ' Rani cut into his tirade. ' They are of no more account than the monkey people ! And they could not touch us if they would. I tell you, only one man comes close, the man who brings meat with him—and he, he is my friend. Listen ! He is coming now ! '

The two animals, each sitting close to the wall that divided their cages, listened intently to the keeper as he walked down the hill which leads from the main road through Whipsnade Park. They heard the scrunch of his heel as he turned down the concrete slope, and both rose when he stood at last outside the iron bars. Rani started to pace backwards and forwards before the bars, and Argh—his eyes watching every movement of the newcomer—followed her example.

' Well, Rani,' said the keeper cheerfully. ' It seems that you've taken kindly to the boy friend we've brought you—and well you might ! He's a fine-looking chap and no mistake ! '

Rani gave him a deep-throated purring growl of welcome and reared with her forefeet against the bars, while Argh stopped his ambling walk and regarded him with interest.

' We are going to turn you two out together,' continued the keeper, ' so mind you behave yourselves. It's a lovely day—or will be when it's a bit older. You've just time for a run together before the public is let in.'

The keeper turned away, and a few minutes later both animals heard the sound of the grating metal behind them as the doors leading from their cages to the pit were gradually opened upwards, like the sluices of a lock.

Argh and Rani stepped eagerly out into the gentle early morning sunshine.

★   ★   ★

That morning was the beginning of a new life for Argh. He still used his cage as sleeping-quarters, but instead of being shut up day and night he was allowed to roam about the cement pit to his heart's content, with Rani as his companion.

The first time a crowd of sightseers came to visit the pair he was nervy and fretful, but Rani, a creature of much common-sense, kept close by his side.

' Do not mind them—they are of no account,' she seemed to whisper to him. ' If you get cross they will keep you shut up in your cage. Look at them ! See how funny they are, with their cubs treading round their feet ! See how they push each other in order to get a better view of us ! The crows and the monkey folk would make less chatter ! '

Argh watched the people above him as they jostled each other, stood on each other's toes, and dug their elbows in each other's ribs in their efforts to see the pair of tigers.

Argh gave a low growl of contempt, and then blew through his nostrils. ' You are right ! ' that snort said. ' They are not worth worrying about.'

He sat down by the big tree trunk that lies in the pit and started to wash his face, to the intense joy of a happy party of visitors who had just arrived from London in a chara-banc.

'Look at that tiger washing his face!' cried one. 'He's as gentle as a cat!'

'Cat my eye!' rejoined a friend. 'Half a minute ago he was hating us like poison—then his young woman said something to him and he shut up. I saw her!'

'Oh Alf, you *have* got an imagination!' chuckled his mother proudly. She looked round to see if the other people had appreciated her son's cleverness sufficiently. A general titter greeted the boy's remark.

From that day Argh minded the crowds as little as Rani, and watched them with great interest. The sound of laughter began to please him, as it pleased Rani, and sometimes he would deliberately court the amused approval of their audience. Had Mr Benson been able to recover Argh now he would have had no difficulty in getting him to perform in public, but that, fortunately for the tiger, was impossible.

In this way a whole year passed—a happy, contented year for both of them. One morning Argh stepped as usual through the door at the back of his cage into the pit to find that he was in sole possession of their playground. Where was Rani?

He called to her, and to his relief heard her answering roar, though it was muffled by the closed door of the cage.

'Aren't you coming out this morning?' he asked. 'Why have they not opened your door?'

'You will see in good time,' replied Rani from within the den. 'Until then you must wait.'

That one answering roar was the only news that he could get out of his mate.

For the next few days Argh did not see the tigress. For many hours he paced up and down outside the blank walls, drawing amused comment from the sight-seers above.

' He doesn't know she's got cubs in there ! '

' Pity he can't read the papers, or he'd know why he's shut out.'

' Let's hope he doesn't eat 'em when they *do* come out.'

But as Argh could not understand the vapourings of the monkey type of man—when his keeper spoke he understood every word—their remarks did not throw any light on the mystery of Rani's absence.

At last the great day came when the authorities considered that it was safe to let him see the cubs. When he was let out into the enclosure he found three writhing balls of claw and fur tumbling over Rani's huge body. To everybody's relief Argh was delighted with them, and the tigers and their family became 'news'. Though Argh did not know it, he figured once more in the press—this time as the father of the first tiger family to be born at Whipsnade.

It was a happy little family. The cubs, born so many miles from their own country, thrived wonderfully, and were a constant source of joy to their parents, and of amusement to the public. But the coming of the cubs had a curious effect upon Argh. Rani and the cubs stayed real enough, but the tiger pit became shadowy and unreal. He was, in imagination, back in the jungle. The crowds that thronged daily above the pit seemed no longer human beings ; they became in his eyes real monkeys chattering among the tree tops.

Though his body remained in the pit his spirit was perpetually roaming through his own jungle, and as the cubs grew older and needed more of their mother's supervision to keep them out of mischief, Argh's power of imagination grew more and more vivid.

One sunny afternoon he lay sunning himself on the rocks, his mind far away. He was wandering through the jungles, while Mr Mainwaring was working in the tin hut that served as his office in that remote part of the forest. He knew that presently his master would call him, and that they would go back to the bungalow for the midday meal.

Soon, as he had expected, Mr Mainwaring's voice sounded through the trees—and suddenly, for the first time for months, the trees were blotted out, and he found himself back in the tiger pit once more ! The jungle had vanished—but the voice remained real !

' Argh, Argh, old man ! Don't you know me ? ' called Mr Mainwaring.

Epilogue

EVEN WHEN Mr Mainwaring at last dragged two unwilling little boys away from the tiger pit Argh was quite unaware of the fact. He was blissfully following his master down a rugged track by the side of a ravine. Every now and again he would turn aside into the jungle and pick up his master's trail at the next bend. Presently the ground became more level, for they were approaching the river. It was his master's custom to cross the water by the rough wooden bridge, but Argh would swim it and then gallop on ahead to the bungalow.

'Isn't there something wrong with that tiger?' called a voice above. 'Look how stiffly he's standing!'

'No! He's going to jump into the pond. He is walking in a funny way all the same. It doesn't look natural to me!'

Argh heard nothing of the voices above him—it was mere monkey talk, or the chatter of the crows. That stiff walk was a gallop, for he was covering the last short stretch of turf between himself and the river.

Splash! He was in! How gloriously cool the water felt. A few strong strokes and he was out on the other side and racing for the bungalow. He flung himself on to his couch in the verandah, as his master came cantering up on his pony!

*   *   *

The newspapers said that Argh had had a fit, and had drowned through falling accidentally into the pool. The water had been drained off as quickly as possible, but when the attendants reached the tiger it was too late to save him.

The spectators had been moved away while the body was being recovered, and one indignant old gentleman wrote to *The Times* and asked what the authorities were doing. As a result a wooden raft has been placed in the pool now, so that it can never happen again.

But the keeper knows better. 'There's been something on his mind ever since he came here,' he said. 'And he's found the best way out.'